The Internet for Molecular Biologists

www.oup.com/pas/internet

Practical Approach Series

For full details of the Practical Approach titles currently available, please go to www.oup.com/pas.

The following titles may be of particular interest:

Essential Cell Biology (2 volume set) (Nos 262 and 263)
Edited by John Davey and Michael Lord

"Recent advances in our understanding of cells have put cell biology at the center of biological and medical research. This two-volume set provides researchers with the information they need to understand and carry out the essential techniques used for studying cells. It covers a wide range of traditional and recently developed techniques and includes the fine detail necessary for immediate application in the laboratory. It is useful both as a compendium of protocols for experienced researchers, and as a valuable guide, for newcomers to the subject."
June 2003 0-19-852764-0 (Pbk set)

Essential Molecular Biology (2 volume set) (Nos 234 and 255)
Edited by Terry Brown

"There is plenty of sound advice, not just on the molecular techniques but on how to be a good scientist in general. Although the advice starts with the basics, it isn't patronising to those experienced in other fields." *Journal of Cell Science*
January 2002 0-19-851091-8 (Pbk set)

Zebrafish (No 261)
Edited by Christiane Nusslein-Volhard and Ralf Dahm

"This book is an excellent tool for researchers working with, or planning to work with Zebrafish. This is information on the background of the techniques used as well as the protocols on how to perform the techniques." *British Society for Development Newsletter.*
September 2002 0-19-963809-8 (Hbk) 0-19-963808-X (Pbk)

Basic Cell Culture (No 254)
Second edition
Edited by John Davis

"I recommend [this book] not only to beginners, but also to more experienced cell culture workers, who will find many new hints and pieces of information, presented in a readily accessible way." *Experimental Physiology*
January 2002 0-19-963853-5 (Pbk)

No. 269

The Internet for Molecular Biologists

A Practical Approach

Edited by

Clare E. Sansom

School of Crystallography,
Birkbeck College, London
WC1E 7HX, UK

Robert M. Horton

Attotron Biosensor Corporation,
2533 N. Carson Street, Box A381
Carson City, NV 89706, USA

OXFORD
UNIVERSITY PRESS

OXFORD
UNIVERSITY PRESS

Great Clarendon Street, Oxford OX2 6DP

It furthers the University's objective of excellence in research, scholarship,
and education by publishing worldwide in

Oxford New York

Auckland Bangkok Buenos Aires Cape Town Chennai
Dar es Salaam Delhi Hong Kong Istanbul Karachi Kolkata
Kuala Lumpur Madrid Melbourne Mexico City Mumbai Nairobi
São Paulo Shanghai Taipei Tokyo Toronto

Oxford is a registered trade mark of Oxford University Press
in the UK and in certain other countries

Published in the United States
by Oxford University Press Inc., New York

A catalogue record for this title is available from the British Library

Library of Congress Cataloging in Publication Data
(Data available)
ISBN 0 19 963887 X (hbk)
        0 19 963888 8 (pbk)

10 9 8 7 6 5 4 3 2 1

Typeset by Newgen Imaging Systems (P), Chennai, India.

Printed and bound in Great Britain by Antony Rowe Ltd., Chippenham, Wiltshire

Preface

The last few decades of the twentieth century will be remembered for two techno-logical revolutions that have already had a profound effect on millions of people's lives. Thanks to communication technologies, and particularly the Internet, we now take immediate access to enormous quantities of information for granted. And thanks to 'the new biology', building on the genome projects, some sci-entists are predicting that cures for the majority of known diseases could be readily available within two generations. These two revolutions are inextricably linked: molecular biology depends on the ready availability of data, and that needs computers and the Internet. There are some surprising parallels between developments in the two technologies. For example, the publicly available gene sequence databases, EMBL and GenBank, are doubling in size approximately every 18 months. According to Moore's Law, which has held since the inven-tion of the silicon chip, 18 months is also the time frame in which computer power is expected to double.

This book aims to help the practitioners of the second revolution—molecular biologists who are more at home at a laboratory bench than in front of a computer keyboard—to use the technology of the first, the Internet, more effectively. To some, this may seem a grandiose ambition, but to others it may appear simply unnecessary. After all, pre-school children are now taught mouse skills. Is Internet literacy not becoming as ubiquitous as literacy itself, at least in Western countries and amongst the younger age groups (including the graduate students who we hope will read this book)?

This parallel with literacy can be used to make a case for the importance of a book like this. Graduate students of molecular biology must be highly literate individuals. Yet they still need to learn how to make sense of the scientific lit-erature: to locate the most relevant papers, read them with understanding, and in due course make their own contributions to that literature. The World Wide Web—the most information rich part of the Internet—has been compared with a worldwide library. This is an apt parallel, but the Web is orders of magnitude larger and richer than any library in the world. It is also far less authoritative, and far harder to explore in depth. There is truth in both of the often-quoted clichés, that only 1% of the Internet is of any conceivable use, and that that 1% contains more information than anyone could absorb in a single lifetime. In this book we offer a guide to that part of the 1% that molecular biologists will find most useful.

A little history

The Internet is a network of interconnecting computer networks. The first of these networks, ARPAnet, was a product of the Cold War. In the late 1950s, the US Defense Department founded the Advanced Research Projects Agency to 'establish a US lead in science and technology applicable to the military'. Developing protocols for communication between distributed computers was seen as an important part of protecting US infrastructure from attack by the Soviet Union. The first four sites at US universities were connected together late in 1969. Two years later, Ray Tomlinson, a Boston programmer, developed a program that allowed people to send messages between computers on ARPAnet, and e-mail was born. He now claims to have forgotten the text of the first e-mail he sent to himself in October 1971. It was certainly more forgettable than the text of the first telegram, sent over a hundred years earlier: 'What hath God wrought!' E-mail is still the most popular Internet application.

By the 1980s, universities throughout North America and Western Europe were connected together via countrywide networks such as the UK's Joint Academic Network (JANet). Molecular biologists were regularly logging in to central servers to run sequence analysis programs, transferring data from one machine to another, and, of course, using e-mail to communicate with their collaborators. The networks were slow and the software was often complex and difficult to use.

It took the invention of the World Wide Web in the early 1990s to turn the Internet into the worldwide cultural phenomenon that it is today. In 1991, Tim Berners-Lee and Robert Caillou, scientists working at CERN in Geneva, invented the Hypertext Transfer Protocol (HTTP) as a way of linking and cross-referencing documents held on different machines. For the first time, the Internet became easy to use, and the rest was history. By 1997, 195 out of 237 countries had Internet access; by 2003, there were over 171 million Internet hosts worldwide. Many professionals, including molecular biologists, cannot now imagine a working life without 'the Net'.

A guide to the guidebook

This book contains three types of information that will help molecular biologists navigate the Internet. First, there are guides to general information sources that biologists need to use increasingly often: bibliographic resources, sequence databases, and phylogeny sites. A second section covers more specialist and subject-specific resources. Separate chapters cover sites concerned with, for example, medical genetics, agricultural biotechnology, and developmental biology. Finally, in a series of connected chapters towards the end of the book, we explain why molecular biologists might want to become active contributors to the Internet as well as passive users of it, and how they can best go about doing so. Scientists and scholars of all disciplines rely on the scientific literature. Jo McEntyre and Barton Trawick of the National Center for Biotechnology Information, the home of Medline, describe how an increasingly electronic scientific

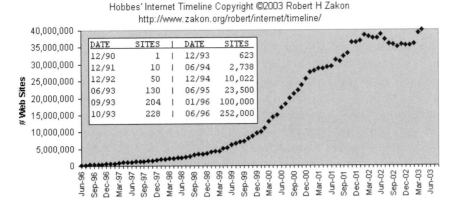

Hobbes' Internet Timeline Copyright ©2003 Robert H Zakon
http://www.zakon.org/robert/internet/timeline/

DATE	SITES		DATE	SITES
12/90	1		12/93	623
12/91	10		06/94	2,738
12/92	50		12/94	10,022
06/93	130		06/95	23,500
09/93	204		01/96	100,000
10/93	228		06/96	252,000

Growth of Internet hosts from Hobbes' Internet Timeline
(http://www.zakon.org/robert/internet/timeline/).

literature is structured and the many tools that are available for searching it. Sequence databases and search tools are covered by Rolf Apweiler and Rodrigo Lopez of the European Bioinformatics Institute, Cambridge, UK. This chapter is a practical guide to the sequence databases. Those readers who need to study the mathematics behind the search tools in detail should also consult a bioinformatics textbook, such as the one by Higgins and Taylor in this series. Korbinian Strimmer (University of Munich, Germany) and David Robertson (University of Oxford, UK) cover Internet tools for determining evolutionary relationships between sequences (phylogenetic analysis) in a similar amount of detail.

Moving on to specialist resources, the NCBI's Kim Pruitt covers medical genetics, and Victoria Carollo of the US Department of Agriculture, agricultural biotechnology. Peter Vize (University of Texas, USA) describes Internet resources for cell and developmental biologists.

A chapter on the use of the Internet to facilitate scientific collaboration bridges the gap between chapters dealing with existing Internet resources and those more concerned with content development. It is written, appropriately enough, by a large and diverse group of authors, headed by Robert Gore-Langton of the EMMES Corporation, Maryland, USA. Chao Lu and James Woodgett of Ontario Cancer Center, Canada, then describe how to 'make friends and influence people' by building a laboratory website. Eric Martz (University of Massachusetts) and Tim Driscoll (molvisions.com, Amherst, USA) describe programs for displaying and manipulating molecular structures that can be used via a Web browser. This aspect of visualization is complemented by Tomaz Amon (Bioanim.com, Ljubljana, Slovenia) who introduces the reader to the capabilities of Virtual Reality Modelling Language (VRML). Finally, my co-editor, Robert Horton, provides a brief introduction to writing web-related programs in the Perl scripting language, and the use of stylesheets for manipulating XML data.

Teaching molecular biology over the net

My own experience at the interface between information technology and molecular biology is largely as a teacher of postgraduate students. At Birkbeck College, London, I, with Professor David Moss and other colleagues, have used many of the technologies described in the third section of this book to develop a completely Internet-based modular M.Sc. course in structural biology. Over 300 students from four continents have taken one or more modules since the programme started in 1996. This was one of the first such courses, but it is not by any means the only one. In the United Kingdom, Manchester University offers an M.Sc. in Bioinformatics on a distance learning basis via the Internet, and the University of Bielefeld in Germany has developed an award-winning short Internet based course in the same subject. In the United States, Massachusetts Institute of Technology is now offering all its teaching materials on the Web, free of charge.

Structural biology is a subject that lends itself very well to Internet-based delivery; it is self-evident that its teaching relies on our ability to provide images of three-dimensional structures that can be manipulated easily. Scripts, written locally or running on Internet-based servers throughout the world, and VRML animations enhance the experience of students on these and similar courses. But studying at a distance can often be a lonely experience. We use a multi-user virtual environment (MUVE) as a synchronous communication tool for tutorials in real time. Students from places as far apart as California and Hong Kong 'meet' to chat together about the course (or anything else that takes their fancy) and to ask questions of the course tutors. In *Rethinking University Teaching* (1993), Laurillard described such an environment as 'the ideal teaching situation of a tutor–student discussion, though at a distance'. At that time, the Web was in its infancy: the Web based environments in use today are ubiquitous and far easier to use.

There are many reasons why students choose to study an Internet-based course. Sarah Lee, from Warwickshire, UK, was one of the first of Birkbeck's Structural Biology M.Sc. graduates. She has explained her motivation: 'Doing a course on the Internet was excellent for me, as I didn't have to go anywhere but to my computer . . . if I hadn't studied in this way, I wouldn't have been able to take an M.Sc. at all.'

A global village. . .?

Students on Birkbeck's Internet courses have sometimes said that, however far from London they live, they nevertheless feel part of the college community. This is a small example of the 'global village' concept developed by Marshall McLuhan decades before the Internet made it a reality. His insight was that an 'electronic nervous system' could integrate the planet so that events in one place could be experienced in real-time in other places. And there can be no event that demonstrated that more profoundly than the terrorist attacks on New York and Washington on 11 September 2001. Within 2–3 h of the atrocities taking place, I had exchanged e-mails with an eyewitness: a client based in Manhattan. Tens, if not hundreds, of millions of people would have had similar experiences to mine, and the wave of empathy that spread around the Net on that day was almost tangible.

But there is another reality that exists alongside the weird, wired world of the Internet generation. It is almost impossible for those of us who have come to rely on the Net to believe the astonishing fact that about half the world's population have never made a telephone call. At a ceremony late in 2001 to mark the centenary of the Nobel Peace Prize, 100 Nobel laureates from many disciplines, including molecular biology, issued the following statement: 'The most profound danger to world peace in the coming years will stem not from the irrational acts of states or individuals but from the legitimate demands of the world's dispossessed.' Perhaps we, in privileged, interconnected countries, can best minimize the likelihood of future atrocities by trying to spread the benefits of the two late twentieth-century technological revolutions throughout the whole world. We don't only owe this to 'the dispossessed': we owe it to ourselves.

Dr Clare Sansom

Contents

4   Agricultural biotechnology *57*

Victoria Carollo

5   Inference and applications of molecular phylogenies: An introductory guide *73*

Korbinian Strimmer and David L. Robertson

10 Virtual reality for biologists   *191*

Tomaz Amon

Protocol list

Abbreviations

AIM	AOL Instant Messenger
API	Application Programming Interfaces
ASP	Active Server Pages
BIRN	Biomedical Informatics Research Network
BLAST	Basic Local Alignment Search Tool
BMJ	British Medical Journal
BRI	Bioscience Resources on the Internet
CDC	Center for Disease Control
CDS	coding sequences
CGAP	Cancer Gene Anatomy Project
CGI	Common Gateway Interface
CluSTr	Clusters of SWISS-PROT
COG	Clusters of Orthologus groups of proteins
COS	Community of Science
CSCW	Computer-supported Cooperative Work
CSS	Cascading Style Sheets
CVW	Collaborative Virtual Workspace
DBRI	Doing Biological Research on the Internet
dbSNP	Database of Single Nucleotide Polymorphisms
DCMI	Dublin core Metadata Initiative
DDBJ	DNA Data Bank of Japan
DFA	Discrete Finite Automation
DSSP	Dictionary of secondary structure in proteins
DU	Diversity University
DV	Digital Video
EBI	European Bioinformatics Institute
EMBL-EBI	European Bioinformatics Institute
EST	Expressed Sequence Tag
FISH	Fluorescent in situ Hybridization
FSSP	families of structurally similar proteins
FTP	File Transfer Protocol
GDR	Genome Database Resource
GMO	Genetically Modified Organisms
GO	Gene Ontology
GPCROB	G-Protein coupled receptor database
GRIN	Germplasm Resources Information Network
HGMD	Human Gene Mutation Database

HSSP	homology-derived secondary structure of protiens
HTML	Hyper Text Mark-up Language
HUGENET	Human Genome Epidemiology Network
IB	Internet Biologists
IM	Instant Messaging
IMGT/HLA	Immunogenetics Human Leukocyte-Associated Antigen Databse
ISI	Institute for Scientific Information
JSP	Java Server-Pages
KEGG	Kyoto Encylopedia of Genes and Genomes
LAN	Local Area Network
LOD	Level of detail
MAGE-ML	Microarray Gene Expression Markup Language
MeSH	Medical subject Heading
MGC	Mammalian Gene Collection
MHC	Major Histocompatibility Complex
ML	Maximum-likelihood
MOO	MU, Object-Oriented
MSPs	Maximal Segment Pairs
MUD	Multi-user Dimension or Domain
MUVE	Multi-user Virtual Envirnoment
NBRF	National Biomedical Research Foundation
NCBF	Non-Covalent Bond Finder
NCBI	National Center for Biotechnology Information
NCI	National Cancer Institute
NGRP	National Genetic Resources Program
NLM	National Library of Medicine
NNTP	Network News Transfer Protocol
OMIM	Online Mendelian Inheritance in Man
PAM	Point Accepted Mutation
PDB	Protein Data Bank
PDQ	Physician Data Query
PE	Protein Explorer
PGP	Pretty Good Privacy
PHP	Personal Hypertext Preprocessor
PKU	Phenylketonuria
PIR	Protien Information Resources
RDF	Resource Description Framework
Refseq	Reference sequences
RTF	Rich Text Format
SCOP	structural classification of proteins
SDB	Society for Developmental Biology
SQL	Structured Query Language
SRS	Sequence retrieval system
SSR	Simple Sequence Repeats
STKE	Signal Transduction Knowledge Environment

STS	Sequence Tagged Sites
SVG	Scalable Vector Graphics
TC	Tentative Consensus
URI	Uniform Resource Identifier
URL	Uniform Resource Locator
USDS-ARS	US Department of Agriculture Agricultural Research Service
VIWB	Virtual Institute of Experimental (wet) Biology
VRML	Virtual Reality Modeling Language
VSNS	Virtual School of Natural Science
XML	Extensible Markup Language
XSL	Extensible Style sheet Language

List of contributors

Tomaz Amon
Director, Center for Scientific
Visualization
University of Ljubljana
Faculty of Electrical Engineering,
Trzaska 25
1000 Ljubljana, Slovenia
tomaz.amon@siol.net

Rolf Apweiler
EMBL Outstation,
The European Bioinformatics
Institute,
Welcome Trust Genome
Campus, Hinxton,
Cambridgeshire,
CB 10 1SD, UK
rolf.apweiler@ebi.ac.uk

David Atherton
Texas A&M University,
Health Science Center,
Cardiovascular Research
Institute
Building 205, 1901 S. 1st St.
Tempe, TX, USA 76504

Victoria Carollo
United States Department of
Agriculture,
Agricultural Research Service,
800 Buchanan Street,
Albany CA 94710, USA
vcarollo@pw.usda.gov

Timothy Driscoll,
Molvisions.com, Waltham,

MA, USA
driscoll@molvisions.com

Robert E. Gore-Langton
The EMMES Corporation,
401 N. Washington St.
Suite 700, Rockville, MD 20850,
USA
rlangton@emmes.com

Robert M. Horton
Attotron Biosensor Corporation,
2533 North Carson Street,
Box A381,
Carson City, NV 89706, USA
rmhorton@attotron.com

Pinar Kondu
Iontek, Inc.,
Meridyen Is Merkezi
Ali Riza Gurcan Cad.,
Cirpici Yolu No. 1/410
Merter 34010
Istanbul, Turkey

Chin Hoon Lau
Lagenda Knowledge Systems Sdn.
Bhd.,
1A, 15th Floor
Komtar Jalan An Fook,
80000 Johor Bahru, Johor

Zev Leifer
New York College of Podiatric
Medicine,
53 East 124th St.
New York, NY 10035, USA
leiferl@ix.netcom.com

Rodrigo Lopez
EMBL Outstation,
The European Bioinformatics
Institute, Welcome Trust Genome
Campus, Hinxton,
Cambridgeshire, CB 10 1SD, UK
Rodrigo.Lopez@ebi.ac.uk

Chao Lu
Microarray Facility,
The Hospital for Sick Children
555 University Avenue
Elm Wing Room 10104
ON Canada M5G 1X8
chao@genet.sickkids.on.ca

Eric Martz
University of Massachusetts,
Amherst, MA, USA
emartz@microbio.umass.edu

Johanna R. McEntyre*
National Center for Biotechnology
Information
National Library of Medicine
National Institutes of Health
Bldg. 45, Room 5an12.
45 Center Drive,
Bethesda, MD 20892,
USA
mcentyre@ncbi.nlm.nih.gov

Kim D. Pruitt
National Center for Biotechnology
Information,
National Institutes of Health,
8600 Rockville Pike,
Bethesda, MD 20894, USA
pruitt@ncbi.nlm.nih.gov

David L. Robertson
2.205 Stopford Building,
University of Manchester,
Oxford Road,
Manchester,
M13 9PT,
UK
david.l.robertson@man.ac.uk

Korbinian Strimmer
Department of Statistics,
University of Munich,
Ludwig Str. 33,
80539 Munich
Germany
strimmer@stat.uni-muenchen.de

Barton W. Trawick
National Center for Biotechnology
Information,
National Library of Medicine,
National Institutes of Health,
Bldg. 45, Room 5an12.
45 Center Drive,
Bethesda, MD 20892,
USA
trawick@ncbi.nlm.nih.gov

Peter D. Vize
Department of Biological
Sciences,
2500 University Drive N.W.,
Calgary, Alberta,
Canada T2N 1N4.
pvize@ucalgary.ca

James R. Woodgett
Ontario Cancer Institute/Princess
Margaret Hospital,
610 University Avenue,
Suite 10-303, Toronto,
ON Canada M5G 2M9
jwoodget@uhnres.utoronto.ca

Chapter 1
Bibliographic databases

Barton W. Trawick and
Johanna R. McEntyre

National Center for Biotechnology Information, National Library of
Medicine, National Institutes of Health, Bethesda,
MD 20892, USA.

Use of the literature is fundamental to the pursuit of all knowledge. Through searching and reading, we learn what our peers are doing, develop a broader perspective on our field of interest, get ideas, and confirm our discoveries. During the course of twentieth century science, 'the literature' has become an expanding knowledge base that represents the collective archive of the work carried out by the international scholarly community. Recent technological advances make an increasing proportion of the literature available electronically (see *Figure 1*). This chapter offers an introductory guide for molecular biologists to stable bibliographic resources that are available over the Internet.

1 General introduction

The term 'bibliographic databases' has traditionally referred to the 'abstracting and indexing services' for the scholarly literature. These services focused on collecting the citation information and abstracts of research articles and making them searchable. Abstracts have been the focus for the creation of bibliographic databases because they summarize the full research article, are small enough to re-key (the only way to capture the information before electronic publishing), store, and search.

However, technological advances over the past decade have expanded the horizons of bibliographic database creation from using abstracts only to using longer pieces of text. Furthermore, the rise in use of the Internet has provided the opportunity to build online, searchable literature databases that are accessible to anyone with an Internet connection.

In response to this opportunity, publishers, libraries, and other information providers have adopted new electronic publishing technologies to develop many forms of online content. These include databases of journal abstracts, full-text

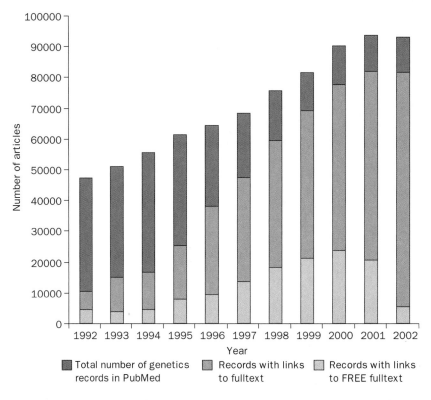

Figure 1 The number of PubMed records that are classified under the general MESH term "Genetics" has grown from around 48,000 in 1992 to 93,000 by the end of 2002. Around 1995, a number of science, technical, and medical (STM) journals began to establish an online presence.[a] Since that time, the proportion of records that provide links to online full-text articles has increased. In 2001 and 2002, around 87% of "Genetics" records have links to online full-text articles; about 25% of these are freely available. (The smaller proportion of free-access articles in 2002 is indicative of the common practice of publishers to delay free access for a period of time after publication.)

[a] Hitchcock S, *et al*. 1996. A survey of STM online journals 1990–95: the calm before the storm.
http://journals.ecs.soton.ac.uk/survey/survey.html

articles, and books, as well as 'internet-only content' in the form of news and summaries.

For the purposes of this chapter, 'bibliographic databases' will be considered as 'any large, stable, collection of primarily text-based information that is available over the Internet'. The chapter will therefore not discuss individual journal titles (although it could be argued that the online collection of articles from a single journal constitutes a small database), nor will it discuss the more popular health and medicine websites (though there are many to choose from). Further, molecular biology sequence databases frequently have some literary or descriptive component; but if their focus is on data rather than text, it will not be discussed here. As electronic publishing is in such a state of flux, the discussion will be limited to only the most stable resources on the web.

Many of these databases require a personal subscription, a library subscription, or a site license, but several of the resources discussed are free to use.

This chapter is divided into *three* sections, based on the following types of bibliographic information described:

1 *Abstracts*: bibliographic databases that contain the abstracts of journal articles plus the citation information (e.g. author names and affiliations, the journal title, volume, and page numbers).
2 *Full-text articles*: there are now several resources available on the web that offer free access to the complete articles from life science journals.
3 *Books and text-rich websites*: some publishers—both traditional and new ones—are now experimenting with the online publication of textbooks, as well as new, information-rich websites.

2 Abstracts

When investigating a new topic area or seeking an update on a known research area, searching an online collection of abstracts of journal articles is often the first approach.

The strategy used to search abstracts databases is central to how successful you will be in finding what you are looking for (or discovering things you did not know you were looking for!). A search query that is too broad will pick up so many abstracts as to be useless, while one too specific might be too limiting for an expansive search. It takes practice to find the right balance and may require the use of Boolean search constructs and techniques such as delimiting the search by restricting it to specific fields, for example, searching only author names, or only article titles. A general introduction to the use of advanced search techniques is summarized in *Box 1*.

Box 1 Tips and tricks for searching bibliographic databases

Boolean searching
Boolean expressions (named in honour of the English mathematician George Boole) allow the user to combine 'AND', 'OR', and 'NOT' operators with specific search terms in order to create a more defined query. In most search engines, operators may be combined (solved in order from left to right), and parentheses '( )' may be employed to clarify terms, group them together, and change the order in which expressions are solved.

Box 1 (*Continued*)

The following expressions:	Will return references that contain:
Watson AND Crick	*at least* both terms 'Watson & Crick'
Watson OR Crick	Either 'Watson' or 'Crick', or 'Watson & Crick'
Watson NOT Crick	'Watson' but not 'Crick'
Wilkins AND Watson AND Crick	*at least* 'Wilkins & Watson & Crick'
(Watson AND Crick) OR Wilkins	'Watson & Crick' or 'Watson, Crick & Wilkins' or 'Wilkins'
Wilkins AND (Watson OR Crick)	'Wilkins & Watson' or 'Wilkins & Crick' or 'Wilkins, Watson, & Crick'
Watson AND Crick NOT Wilkins	*at least* both terms 'Watson & Crick' with no occurrences of 'Wilkins'

Limiting searches to fields

In addition to using Boolean expressions to define a more specific output, it is also possible to limit individual terms to fields. For example, the term 'Crick' may be limited to 'Author Name' and 'Nature' may be limited to 'Journal Name'. The number of possible fields that are in a given database may vary, but at a minimum usually include: Author Name, Author Affiliation, Journal Name, Article Title, Publication Date, Page Number, Issue, and Volume.

History functions

The results of two or more searches can be combined to form a third output, or additional terms may be added to results from previous searches through the use of 'history' functions. This can be particularly useful for reducing large search results into smaller, more focused ones or for combining several different terms with a single common term. For example, independent queries for 'cancer', 'DNA repair', '1995', 'Vogelstein', 'human', and 'mouse' could be used in various permutations and combinations (linked by Boolean expressions) to form new queries. Additionally, some bibliographic databases can be customized so that useful queries can be stored for future use.

For an example of how these search techniques can be combined to search PubMed, see *Protocol 1*.

2.1 Databases—in all their forms

There are several abstracting and indexing services available online, many of which require a subscription. There is considerable content overlap among the major bibliographic databases, and for this reason your library is unlikely to subscribe to all of them.

When considering the use of any of these databases, it is important to make a distinction between the database itself (i.e. the physical collection of abstracts) and the access route into the information. Several of the large databases can be accessed from more than one place, because the owners of the data (i.e. the abstracts collection) lease or sell their data, or have allowed service providers to furnish a portal to the information.

Table 1 lists the major abstracts databases for molecular biology, along with the owner of the database and a list of access points into the data. MEDLINE, for example, is one of the most widely used abstracts databases. Some of the abstracts of MEDLINE can be distributed freely (those under copyright require permission to reproduce them), so many organizations have developed clones or interfaces to MEDLINE that generate alternative portals to the same information. This can

Table 1 Abstracts databases

Resource	Produced by	Examples of access	Free access*	URLs
PubMed/MEDLINE	The National Library of Medicine (NLM)	PubMed BioMedNet Ovid BIDS	Yes Yes No Yes	http://www.pubmed.gov http://research.bmn.com/medline http://www.ovid.com/ http://www.bids.ac.uk/
ISI Citation Database (Web of Science)	Institute for Scientific Information (ISI)	Web of Science	No	http://www.isinet.com/isi/journals/
Current Contents®	Institute for Scientific Information (ISI)	Current Contents Connect Ovid	No No	http://www.isinet.com/isi/journals/ http://www.ovid.com/
BIOSIS Previews® (comprising biological abstracts and biological abstracts/RMM®)	BIOSIS	BIOSIS Ovid	No No	http://www.biosis.org/ http://www.ovid.com/
Pascal	Institut de l'Information Scientifique et Technique	BIDS	Yes	http://www.bids.ac.uk/
EMBASE	Elsevier Science	EMBASE.com Ovid	No No	http://www.embase.com/ http://www.ovid.com/
The Cochrane Reviews (abstracts)	The Cochrane Library	The Cochrane Library	Yes	http://www.update-software.com/abstracts/crgindex.htm

* In cases where access to the database is not free, consult your library for subscription information.

provide a useful addition to a publisher's website, or it may produce an interface in a language other than English.

2.1.1 The databases

2.1.1.1 *PubMed/MEDLINE*

PubMed was developed at the National Center for Biotechnology Information (NCBI), within the National Library of Medicine (NLM), USA. It encompasses the over 12 million abstracts in MEDLINE, and currently covers about 4000 biomedical journals, dating back to 1966. MEDLINE abstracts have a controlled vocabulary associated with them known as Medical Subject Heading (MeSH) terms. Several terms are assigned to each MEDLINE abstract, and are used for indexing articles to provide a consistent way to retrieve information.

As well as enabling abstract searches (e.g. see *Protocol 1*), PubMed offers the following additional functions:

1 Links to biological sequence information, including data such as GenBank protein and nucleotide sequences, and macromolecular structures.

2 Links to the full-text of journal articles (about 4000 journals are currently linked in this way). Whether the full text can be viewed without purchasing the journal depends on the journal policy (see section below on full-text articles).

3 Links to 'Related articles'. For each abstract, similar articles in the database have been identified, based on a statistical analysis of words and phrases found in the abstract text. This is an easy way to expand on a PubMed search when a useful abstract has been found.

4 Links to resources outside of the NLM. The 'LinkOut' feature allows other providers of information, such as organism-specific databases like FlyBase, to link to related abstracts.

5 Links to textbooks. A new collaborative project at the NCBI is linking the content of textbooks to PubMed abstracts to serve as background information (see Section 4.2).

PubMed is primarily a biomedical database that historically has not collected abstracts from non-medical areas of molecular biology. However, more recently, the scope of PubMed has widened to include coverage of those areas, such as the plant sciences. PubMed does have the significant advantage that it can be used free-of-charge from anywhere in the world.

Protocol 1

Using PubMed

The PubMed page (*Figure 2(a)*) consists of the following: (1) A sidebar that contains links to PubMed information and services; (2) A query box for entering search terms; (3) A feature bar that contains links for advanced searching; and (4) Links to other integrated molecular biology databases.

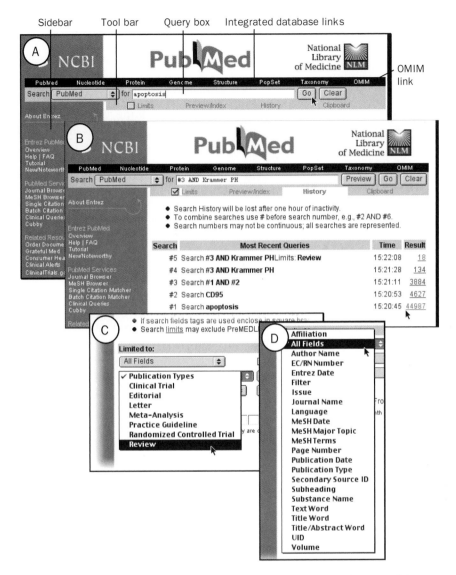

Figure 2 The PubMed search interface. (a) The PubMed page has a sidebar that links to related services and information, links to integrated databases, a query box for entering search terms, and a tool bar that contains links for advanced searching. (b) Previous searches can be viewed using the history feature. Searches can be combined through use of search numbers and Boolean operators. (c) The 'Limits' feature allows searches to be constrained to various information fields, such as Author Name or Review Article. (d) The field restrictions are found only in the 'All Fields' pull-down menu.

Protocol 1 continued

As an example of how a PubMed search can be conducted, we will look for review articles on CD95 (or Fas, a lymphocyte receptor) and apoptosis, written by P. H. Krammer.

Method

1 Enter the search term 'apoptosis' in the query box and click the 'Go' button to the right of the box to initiate the search (*Figure 2(a)*). Conducting a search using a broad term without any field restrictions usually returns a large number of hits; in this case, more than 70,000 citations are found.

2 PubMed retains the most recent search term in the query box. Click the 'Clear' button to the right of the query box to remove the previous search term ('apoptosis') and replace it with the new search term 'CD95' and click 'Go'. The search for 'CD95' returns several thousand references.

3 Click on the 'History' tab located in the feature bar. The results of the two previous searches are now displayed chronologically in a numbered list. These results may be reviewed individually by clicking on the number of returns for each query in the 'Results' column (*Figure 2(b)*).

4 Searches stored in History may be combined using Boolean operators (see *Box 1*) to form a new search. Search for references that contain both 'apoptosis' and 'CD95' by typing '#1 AND #2' in the query box and clicking on the 'Preview' button. Selecting 'Preview' will display the search results in History summary format, rather than listing each article found. (Note: PubMed is case-insensitive for search terms but case-sensitive for Boolean operators: make sure 'AND' is in capitals.)

5 New queries may be combined with previous searches. To find references associated with[a] the name P. H. Krammer, type '#3 AND krammer ph' into the query box (*Figure 2(b)*) and click 'Go'.

6 Queries can be limited to various fields such as: journal name, author name, title word, MESH term, publications type, publication date (or date range), or language. To employ limits in PubMed, click on the 'Limits' tab in the feature bar. Limit the 'Publication Types' pull-down menu to 'Review' (*Figure 2(c)*). Check that the search term 'krammer ph' is still present in the query box and click 'Go'. The result displays a summary view of all review articles associated with[a] P. H. Krammer that contain the query terms 'apoptosis' and 'CD95'.

7 Limits will remain in effect for subsequent searches unless they are deselected by clicking the check box to the left of the 'Limits' tab in the feature bar. Additional limits are located in the 'All Fields' pull-down menu (*Figure 2(d)*). For a more comprehensive account on advanced searching of PubMed, consult the help documentation, listed in the sidebar on the PubMed homepage.

[a] When searching for authors by last name only, rather than for abstracts that merely cite the author's last name, the search should be carried out with the field limited to

Protocol 1 continued

Author Name. For example, a search for 'Crick' without any field limits will return abstracts that contain the word 'Crick', such as in the term 'Watson–Crick base pair'. Similarly, a search for articles in the journal 'Cell' needs to be executed with the field limited to Journal Name, otherwise the results will list any abstract that contains the word 'cell'.

2.1.1.2 *Web of science*

The Institute for Scientific Information (ISI) produces the 'Web of Science'—an interface to the ISI Citation Database that contains more than 5300 scientific articles, dating from 1980, that is updated weekly. The Web of Science is a subscription-based service, available from many (but not all) university libraries. The Web of Science shares some features with PubMed, such as links to biological sequence information and full-text articles, but also has some that are unique:

1 Links to related articles. The way in which related articles are calculated in the Web of Science differs from the related articles of PubMed. In Web of Science, the list of records related to a given article consists of papers that cite at least one source also listed in the original (parent) article, with the source that has most common citations listed first.
2 Links to (i) the Derwent Innovations Index, a patent database; (ii) BIOSIS Previews, a database of references to primary journal literature, meetings, and books; (iii) ISI Chemistry Server, for newly reported structural chemistry.

For many molecular biologists, one of the most valuable attributes of the Web of Science comes from the use of the citations associated with each abstract. Through the references cited within an article, it is possible to:

(a) View the abstracts of all articles cited in the original (parent) article,
(b) Find all articles published, since the original (parent) article, that have cited it, and
(c) Find all the articles that have cited a particular author.

2.1.1.3 *Current contents*

The Web of Science also interfaces with the ISI Current Contents databases, for which a subscription is required. Current Contents used to be a paper publication, distributed weekly and consisting of the contents of recently published journals, divided into broad subject categories, such as the Life Sciences (coverage of about 1400 journals). The Current Contents database can be searched, abstracts of articles found can be viewed, and from there the table of contents of the journal issue can be displayed and browsed.

2.1.1.4 *EMBASE*

EMBASE (1974–present) is a bibliographic database produced by Elsevier that covers over 4000 journals in the biomedical and pharmacological sciences. Its online presence now incorporates selected MEDLINE records, thus increasing the scope and scale of EMBASE to over 13 million abstracts. Like PubMed and Web of Science, EMBASE has links from appropriate abstracts to selected full-text articles and gene sequence information. EMBASE is available by library subscription only.

2.1.1.5 *The Cochrane Abstracts*

The Cochrane Reviews is a collection of reports that collate and summarize published health care evidence on a wide range of medical disorders and conditions. The target audience is very broad, ranging from those receiving care, to those responsible for research, teaching, funding, and administration of health care at all levels.

   The reports are written and maintained by international panels of clinicians, who are organized into groups on the basis of area of expertise. There are currently about 50 Collaborative Review Groups that cover areas such as breast cancer, schizophrenia, HIV/AIDS, and tobacco addiction. Once a review is written, it is checked regularly and updated as needed.

   While a subscription is required to access the full Cochrane Reviews, anyone can browse or search the Cochrane Abstracts without charge. The abstracts alone are quite substantial (usually about 300–500 words). They outline the background for the study, the source data, search strategy, and criteria for inclusion, and then state the results and conclusions. The Cochrane Abstracts, while more focused on clinical trials and therapies than basic molecular biology, are a high-quality and useful adjunct for those who work on molecular biology problems with clinical applications.

2.1.1.6 *BIOSIS Previews*

BIOSIS Previews is made up of two databases: Biological Abstracts, which contains about 12 million records from more than 5000 journals, and Biological Abstracts/RRM, which covers reports, reviews, and meetings—information not formally published in scientific research journals. This includes references to items from meetings, symposia, and workshops, review articles, books, book chapters, software, and US patents related to the life sciences. It covers the biological sciences, from biochemistry to zoology, and is available by subscription only.

2.1.2 Access providers: BIDS and Ovid

BIDS and Ovid are companies that aggregate databases created by other organizations into convenient packages for libraries to use. BIDS may be the best-known bibliographic service for academics in the United Kingdom and Ireland. It provides access to a number of databases, some of which are freely available, and links to full text articles via Ingenta Journals (see Section 3). Many databases and services formerly provided by BIDS, including Medline, are now provided free to UK academics via the ISI's Web of Knowledge interface (`http://wok.mimas.ac.uk`).

Your library may also use Ovid as a provider of several bibliographic databases, including BIOSIS Previews®, Current Contents®, EMBASE (Excerpta Medica Database), and MEDLINE, among others.

Database aggregators often implement databases in their own way, so the interface for searching the databases may have several features that differ from the implementations of other providers.

3 Full text of research articles

Most of the databases described above concern abstracts of published research articles. Although not considered traditionally as bibliographic information, no discussion of online text resources would be complete without considering the increasing availability of full-text articles.

Several thousand molecular biology journals are now available in electronic form (*Figure 1*)—most are online counterparts to paper journals, but some are online-only publications (see *Box 2* for a summary of the advantages of online articles over articles printed on paper). All can be viewed via a web browser, providing that, with a few exceptions such as the *Journal of Clinical Investigation*, you have a subscription. However, more recently, some journals have made articles from back issues freely available, and new publishing ventures that offer free access to articles are emerging (see *Table 2*).

Box 2 Advantages of online journals over paper journals

Searchable content
Articles in digital format may be searched for words and phrases. Most bibliographic databases provide a search engine that allows for content matching across all entries. Once an article has been obtained, the 'Find' feature in your web browser can be used to search within the article for specific words and phrases.

Hypertext links
Online articles displayed in HTML can exploit hypertext linking to create connections between related content. Links can be made from references cited in the text, to its listing in the bibliography, or to external information sources such as PubMed abstracts, referenced citations, errata, sequence information, macromolecular structures (PDB files), or even the author's home page.

Multimedia
The content of traditional printed journals is restricted to what can be presented on paper. However, online journals are able to 'add value' to articles with movies, audio, and the inclusion of large data sets (an entire genome sequence, for instance). Additionally, use of color figures does not generally represent a higher publishing cost for online journals as it does for print journals.

Box 2 (*Continued*)

Accessibility

Electronic articles can be accessed over the Internet without visiting a library. This is particularly useful for those in remote locations. Downloaded electronic articles can be stored on your personal computer.

Flexible publishing model

Some scientific journals make their online content available before the printed copy. Some journals even provide a 'rolling model' of publication where articles are accessible online as soon as they are accepted for publication. Manuscript submission, online peer review, and access of electronic content may be provided by some journals through the Internet.

Table 2 Online full-text journals

Resource	Produced by	No. of journals[a]	Free access[b]	URL
Science Direct	Elsevier Science	1100	No	http://www.sciencedirect.com/
Link	Springer-Verlag	500	No	http://link.springer.de/
Interscience	Wiley	300	No	http://www.interscience.wiley.com/
BioMed Central	Current Science	130	Yes	http://www.biomedcentral.com/
Society and small publisher online journals[c]	Highwire Press	340	Some	http://highwire.stanford.edu/ and individual journal URLs
PubMed Central[d]	The National Library of Medicine	150	Yes	http://www.pubmedcentral.gov

[a] Journal figures given in round numbers. Figure represents the total number of journals in each resource; not all of these may be life science journals.

[b] In cases where access to the database is not free, consult your library for subscription information.

[c] HighWire Press enables small publishers to make their journal content available online. It is not the publisher of these journals.

[d] PubMed Central is an active archive for journal content; it is not a publisher.

3.1 Access to the full text of research articles

In the absence of a search engine that indexes a good proportion of full-text life science journals, the best route to finding full-text articles is not always obvious. As mentioned above, access to an article is only possible if you or your library has a subscription, or if the article is made freely available. Below we outline the most common and useful routes to online journal articles.

3.1.1 Access through abstracts databases

Most of the databases listed in the previous section can make links between abstracts and the corresponding online full-text article. There will be a link that leads the user seamlessly to the article if the following is true:

1 The journal (more specifically, the journal issue) is published online.
2 The publisher of the journal has agreed with the database to make the article available via this route.
3 You or your library subscribes to the journal, or the publisher makes the article freely available.

For example, a search of an abstracts database will result in a list of 'hits' consisting of the citation information for each article retrieved by the query (see *Box 1*). If the abstract satisfies points (1) and (2) above, then there will be a link to the journal publisher's website (this may only become apparent when viewing the complete abstract rather than the citation information). Clicking on this link will take you to the full-text of the article if point (3) is satisfied. Many of the freely available articles can be found in this way, and, as an example of scope of access, about 4000 journals currently have links from PubMed abstracts to their respective articles on the publisher's site.

3.1.2 Access from publisher sites

Many publishers do not collaborate with all bibliographic databases to allow access to their journals, and the most conservative may only allow access to their journals by logging-on directly to their own website. In these cases, the only way to access the full-text is through your library's interface to the journal, or by a direct visit to the journals' website, if you hold a personal subscription. Here we will list some of the most significant places where there is a collection of full-text articles (see also *Table 2*).

3.1.2.1 *HighWire Press*

HighWire Press works with scientific societies and publishers to create online counterparts to their print journals. There are currently over 340 journals that are available at HighWire, of which about 150 now offer free access to back issues of the journal.

The period of time after which the article becomes freely available depends on the policy of the journal. Some journals, such as *British Medical Journal* (*BMJ*) have an immediate free-access policy (i.e. anyone can look at the most current version of *BMJ*). However, most HighWire journals operate under a delayed-release policy for free full-text articles, ranging from 2 months to 5 years, with most opting for a 1–2 year delay. In total, there are now (Spring, 2001) around 250,000 free articles available. HighWire allows a basic search across all the journals they collaborate with, although the free articles are not clearly delineated.

3.1.2.2 *Individual publishers*

Many publishers have developed their own online interfaces to their journal databases. Some of the largest of these are listed in *Table 2*, although there are

many smaller collections. For all these sites there is usually some free introductory content, but the journal content is almost always available only on a
subscription basis.

3.1.3 Archives for full-text articles

Publishing journals online is still a relatively new enterprise. Now that there is
a substantial volume of information available over the Internet, the question
of how to effectively archive the data and make the best use of the electronic
medium for searching and linking becomes obvious.

A recent initiative called PubMed Central, based at the National Library of
Medicine USA, is aimed towards creating an archive for full-text life science journal articles that can be browsed and cross-searched freely. The idea is that any
journal article available via the PubMed Central site can be viewed by anyone
with an Internet connection from anywhere in the world.

Currently, about 150 journals are making their content available via PubMed
Central, and though small at present, the potential of this kind of initiative for
the future makes PubMed Central worthy of mention.

4 Books and text-rich websites

While books have been less evident than journals in making the transition from
paper to electronic form, a few online texts do exist, although most require
a subscription or site license. A growing trend is for books to have associated
websites for further information and corrections (as this book has*). These are
usually listed prominently in the book. Furthermore, as biological content on the
Internet evolves, so do content-rich websites that do not fit into any traditional
bibliographic mold; this category of bibliographic resource is not well-defined, so
here we will discuss just two of the larger and more stable resources (see *Table 3*
for URL).

4.1 Online Mendelian inheritance in man

Online Mendelian Inheritance in Man (OMIM) is a catalogue of human genes and
genetic disorders (see *Table 3* for URL). It now contains about 15,000 records, and is
authored and edited by Dr Victor A. McKusick and his colleagues at Johns Hopkins
and elsewhere. The online version has been developed by the NCBI.

The OMIM database is usually searched using the name of a genetic disorder
or the name of a gene to retrieve records, and it is possible to use Boolean search
constructs as well as field limitations, such as chromosome number (see *Box 1*).

Table 3 Online books

Resource	URL
Online Mendelian Inheritance in Man	`http://www.ncbi.nlm.nih.gov/entrez/query.fcgi?db=OMIM`
Online books at NCBI	`http://www.ncbi.nlm.nih.gov/entrez/query.fcgi?db=Books`

* http://www.oup.com/pas/internet

OMIM does not contain any figures or graphics, but it does have links to PubMed, gene, and protein information. OMIM is now one of the databases integrated with PubMed, and can be accessed for searching by clicking the OMIM link on the PubMed search page (see *Figure 2*).

4.2 Online books

A project to put biomedical textbooks online, make them searchable, and integrate them with PubMed and other data resources has recently begun at the NCBI (see *Table 3* for URL). There are currently about 24 books participating in the project, which broadly cover the subject areas of basic molecular and cell biology and genetics (*Figure 3*); more books are set to become available in the near future.

The book collection may be searched directly, using a similar interface to PubMed. In addition, all PubMed abstracts have a 'Books' link; clicking on this link brings up a facsimile of the abstract with hyperlinked terms and phrases that lead to the most relevant sections of the book(s) for the linked phrase. PubMed abstracts are rich in information, but they do not explain the terms or concepts used, so linking abstracts to books as background information may help address this shortfall. The quantity and subject area of hyperlinked phrases in an abstract will depend on how much the content of the abstract overlaps with that of the books available.

While the complete contents of the book are free to use in this way, for some books it is not possible to navigate across the whole book content, from chapter to chapter. In these cases, access is limited to 'stand-alone' chapters or sections.

4.3 Text-rich websites: a word of caution

Any web search engine can also be used to search for molecular biology information. Many publishers, biotech companies, research labs, teachers, and others display information that can be browsed freely.

Information found in this way should be carefully evaluated. Be aware that anyone can publish almost anything on the Internet, so a key factor in assessing the validity of any information found is the reliability of its source. It is important to assess what qualifies the individual or organization to publish the information, and what their motivation for doing so has been. As with any literature search, the information found should be cross-checked and critically evaluated before believing.

5 Summary

Bibliographic information on the Internet for molecular biologists continues to grow. This chapter must really be considered a snapshot, serving as an introduction to the potential for exploring online literature resources. For this reason we have chosen to discuss only the most stable of resources, and have not discussed the specific use of any one search interface. The websites and databases discussed undergo constant evolution, and new resources are continually launched and developed. The Internet moves faster than the print world; we hope that this chapter will at least be in the same race for some time!

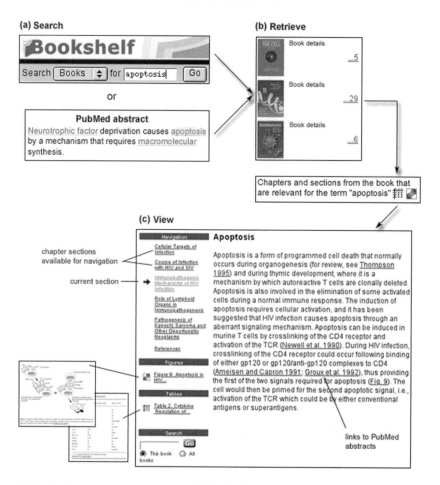

Figure 3 How to access the books at NCBI. (a) All books can be searched directly from the books homepage (see *Table 3* for URL), as well as indirectly, through hyperlinked phrases in PubMed abstracts. Each PubMed abstract obtained by searching PubMed has a 'Books' link. Clicking on this link displays the same abstract with some hyperlinked phrases, as shown here. (b) Executing a books search or clicking on a hyperlink within a PubMed abstract displays a summary list of books in which that term is found. The number on the right indicates the number of book sections that are relevant for the term. This link leads to a book-specific list of sections, figures, and tables. Figures and tables are indicated by the icons shown. (When less than 20 relevant book sections are found, the book summary step is omitted.) (c) The section, table or figure titles lead to the book content. The books are displayed as one chapter section per page, and it is possible to navigate around a minimum of one chapter at a time. The books contain links to the figures and tables of the book, PubMed abstracts, and in the future will be more extensively linked to molecular biology information.

Acknowledgements

We would like to thank Kathi Canese and Edwin Sequeira for carefully reading this manuscript.

Chapter 2

Sequence databases and analysis sites

Rolf Apweiler and Rodrigo Lopez

EMBL Outstation, The European Bioinformatics Institute,
Wellcome Trust Genome Campus, Hinxton, Cambridgeshire,
CB10 1SD, UK.

1 Introduction

Recent years have seen an explosive growth in biological data, which is often not published anymore in a conventional sense, but deposited in a database. Sequence data from mega-sequencing projects may not even be linked to a conventional publication. This trend and the need for computational analyses of the data make databases and database search tools essential for biological research.

Here we introduce the most important molecular biology databases and search tools on the Internet. Of special importance are the sequence databases and sequence similarity search tools, which provide comprehensive sources of information on nucleotide and protein sequences. However, we also describe many other important databases and analysis tools for the life scientist.

The chapter is organized in two main sections. The first gives an overview of groups of databases dealing with:

- nucleic acid data
- protein data
- proteomics related data
- protein families, domains, and functional sites
- structural data.

The second section describes the Entrez and Sequence Retrieval System (SRS) approach to integrate access to these databases, and some recent developments in sequence similarity searches. The URLs of the databases and sites mentioned in this chapter are shown in *Table 1*.

Table 1 The URLs of databases and sites mentioned in this chapter

Database or site	URL
Aarhus/Ghent-2DPAGE	`http://biobase.dk/cgi-bin/celis/`
Acnuc	`http://pbil.univ-lyon1.fr/databases/acnuc.html`
Amos' links	`http://www.expasy.ch/alinks.html`
BLOCKS	`http://www.blocks.fhcrc.org/`
CATH	`http://www.biochem.ucl.ac.uk/bsm/cath/`
CCDC	`http://www.ccdc.cam.ac.uk/`
CluSTr	`http://www.ebi.ac.uk/clustr`
COG	`http://www.ncbi.nlm.nih.gov/COG`
DBGET	`http://www.genome.ad.jp/dbget/dbget.html`
DDBJ	`http://www.ddbj.nig.ac.jp`
DSSP	`http://www.sander.ebi.ac.uk/dssp/,`
EBI	`http://www.ebi.ac.uk`
EBI Blast server	`http://www.ebi.ac.uk/blast2/`
EBI SRS server	`http://srs.ebi.ac.uk`
EMBL Nucleotide Sequence Database	`http://www.ebi.ac.uk/embl/`
Entrez	`http://www.ncbi.nlm.nih.gov/Entrez/`
FSSP	`http://www.ebi.ac.uk/dali/fssp/`
GO	`http://www.geneontology.org`
GPCRDB	`http://www.gpcr.org/7tm/`
HSSP	`http://www.sander.ebi.ac.uk/hssp`
InterPro	`http://www.ebi.ac.uk/interpro/`
KEGG	`http://www.genome.ad.jp/kegg/`
MEROPS	`http://www.merops.sanger.ac.uk`
NCBI	`http://www.ncbi.nlm.nih.gov`
NCBI Blast server	`http://www.ncbi.nlm.nih.gov/BLAST/`
PDB	`http://www.rcsb.org/pdb/`
PIR	`http://www-nbrf.georgetown.edu/`
Pfam	`http://www.sanger.ac.uk/Pfam/`
PRINTS	`http://www.biochem.ucl.ac.uk/bsm/dbbrowser/PRINTS/PRINTS.html`
ProDom	`http://www.toulouse.inra.fr/prodom.html`
PROSITE	`http://www.expasy.ch/prosite/`
Proteome Analysis Database	`http://www.ebi.ac.uk/proteome/`
ProtoMap	`http://www.protomap.cornell.edu/`
SCOP	`http://scop.mrc-lmb.cam.ac.uk/scop/`
SMART	`http://SMART.embl-heidelberg.de`
SP_TR_NRDB	`ftp://ftp.ebi.ac.uk/pub/databases/sp_tr_nrdb`
SWISS-2DPAGE	`http://www.expasy.ch/ch2d/`
SWISS-PROT	`http://www.expasy.ch/sprot/` `http://www.ebi.ac.uk/swissprot/`
SYSTERS	`http://www.dkfz-heidelberg.de/tbi/services/cluster/systersform`
WIT	`http://wit.mcs.anl.gov/WIT2/`

2 Databases

2.1 Nucleotide sequence databases

The International Nucleotide Sequence Database Collaboration is a joint effort of the European Bioinformatics Institute, the DNA Data Bank of Japan, and the US National Center for Biotechnology Information. In Europe, the vast majority of the nucleotide sequence data produced is collected, organized, and distributed by the EMBL Nucleotide Sequence Database (1) located at the EBI, Cambridge, UK. The nucleotide sequence databases are data repositories, accepting nucleic acid sequence data from the community and making it freely available. The databases strive for completeness, with the aim of recording and making available every publicly known nucleic acid sequence.

EMBL, GenBank, and DDBJ automatically update each other every 24 h with new or updated sequences. The result is that they contain the same information, but stored in different formats. Each entry in a database must have a unique identifier, which is a string of letters and numbers corresponding to that record. This unique identifier, known as the Accession Number, can be quoted in the scientific literature, as it will never change. As the Accession Number is permanent, another code is used to indicate the number of changes that a particular sequence has undergone. This code is known as the Sequence Version and is composed of the Accession Number followed by a period and a number indicating which version is at hand. Both the unique identifier and the version number should be used, when referring to records in a nucleotide sequence database.

Since their conception in the1980s, the nucleic acid sequence databases have experienced constant exponential growth reflecting advances in sequencing technology. In May 2003, Release 25 of the DDBJ/EMBL/GenBank Nucleotide Sequence Database had more than 32 billion nucleotides in more than 25 million entries. These archives currently double in size about every eighteen months. Electronic bulk submissions from the major sequencing centres overshadow all other input.

2.2 Protein sequence databases

The protein sequence databases (2) are the most comprehensive source of information on proteins. It is necessary to distinguish between universal databases covering proteins from all species and specialized data collections storing information about specific families or groups of proteins, or about the proteins of a specific organism. Two categories of universal protein sequence databases can be discerned: simple archives of sequence data; and annotated databases where additional information has been added to the sequence record. We describe the Protein Information Resource (PIR), the oldest protein sequence database; SWISS-PROT, an annotated universal sequence database; and TrEMBL, the supplement of SWISS-PROT, which can be classified as a computer-annotated sequence repository. We then consider issues of completeness and redundancy, and some examples of specialized protein sequence collections.

2.2.1 The Protein Information Resource

The National Biomedical Research Foundation (NBRF) established PIR (3) in 1984 as a successor of the original NBRF Protein Sequence Database. Since 1988 the database has been maintained by PIR-International, a collaboration between the NBRF, the Munich Information Center for Protein Sequences (MIPS), and the Japan International Protein Information Database (JIPID).

The PIR-PSD (PIR Protein Sequence Database) release 77.02 (July 2003) contained 283,329 entries partitioned into four sections. Entries in PIR1 are fully classified by superfamily assignment, fully annotated and fully merged with respect to other entries in PIR1. The annotation content as well as the level of redundancy reduction varies in PIR2 entries. Many entries in PIR2 are merged, classified, and annotated. Entries in PIR3 are not classified, merged, or annotated. PIR3 serves as a temporary buffer for new entries. PIR4 was created to include sequences identified as not naturally occurring or expressed, such as known pseudogenes, unexpressed ORFs, synthetic sequences, and non-naturally occurring fusion, crossover, or frameshift mutations. PIR-International and its partners, EBI (European Bioinformatics Institute) and SIB (Swiss Institute of Bioinformatics), have been awarded an NIH grant to produce a single worldwide database of protein sequence and function, UniProt, by unifying the PIR, SWISS-PROT and TrEMBL database activities.

2.2.2 SWISS-PROT

SWISS-PROT (4) is an annotated protein sequence database established in 1986 and maintained collaboratively by the Swiss Institute of Bioinformatics (SIB) and the EBI. It strives to provide a high level of annotation, a minimal level of redundancy, a high level of integration with other biomolecular databases, and extensive external documentation. Each entry in SWISS-PROT is thoroughly analysed and annotated by biologists to ensure a high standard of annotation and maintain the quality of the database (5). SWISS-PROT contains data that originates from a wide variety of organisms; in August 2003 the database release 41.20 contained 132,675 annotated sequence entries from more than 8000 different species.

In SWISS-PROT two classes of data can be distinguished: the core data and the annotation. For each sequence entry the core data consists of the sequence data; the citation information (bibliographical references) and the taxonomic data (description of the biological source of the protein), while the annotation describes:

- Function(s) of the protein
- Post-translational modification (carbohydrates, phosphorylation, acetylation, GPI-anchor, etc.)
- Domains and sites (calcium binding regions, ATP-binding sites, zinc fingers, homeobox, kringle, etc.)
- Secondary structure

- Quaternary structure (homodimer, heterotrimer, etc.)
- Similarities to other proteins
- Disease(s) associated with deficiencie(s) in the protein
- Sequence conflicts, variants, etc.

In SWISS-PROT annotation is mainly found in the comment lines (CC), in the feature table (FT) and in the keyword lines (KW). Most comments are classified by 'topics'; this approach permits the easy retrieval of specific categories of data from the database.

Many sequence databases contain, for a given protein sequence, separate entries corresponding to different literature reports. SWISS-PROT tries as much as possible to merge all these data to minimize the redundancy of the database. Any conflicts between various sequencing reports are indicated in the FT of the corresponding entry.

SWISS-PROT provides the users of biomolecular databases with a high degree of integration between different data collections. Release 41.20 is referenced to 52 different databases through the use of more than 118,000 pointers to information found in data collections other than SWISS-PROT.

2.2.3 TrEMBL

Maintaining the high quality of SWISS-PROT is a time-consuming process that involves extensive sequence analysis and detailed curation by expert annotators. It is the rate-limiting step in the production of the database. A supplement to SWISS-PROT was created in 1996, since it is vital to make new sequences available as quickly as possible. This supplement, Translation of EMBL nucleotide sequence database (TrEMBL), consists of computer-annotated entries derived from the translation of all coding sequences (CDS) in the EMBL nucleotide sequence database, except for those already included in SWISS-PROT. TrEMBL is split into two main sections, SP-TrEMBL and REM-TrEMBL. SWISS-PROT TrEMBL (SP-TrEMBL) contains the entries (944,868 in release 24 June 2003), which should be eventually incorporated into SWISS-PROT. REMaining TrEMBL (REM-TrEMBL) contains the entries that will not get included in SWISS-PROT (98,372 in release 24 June 2003).

TrEMBL follows the SWISS-PROT format and conventions as closely as possible. The production of TrEMBL starts with the translation of CDS in the EMBL nucleotide sequence database. At this stage all annotation in a TrEMBL entry comes from the corresponding EMBL entry. The first post-processing step is to reduce redundancy (6) by merging separate entries corresponding to different literature reports. If conflicts exist between various sequencing reports, they are indicated in the feature table of the corresponding entry. This stringent requirement of minimal redundancy applies equally to SWISS-PROT + TrEMBL. The second post-processing step is the automated enhancement of the TrEMBL annotation to bring TrEMBL entries closer to SWISS-PROT standard (7, 8).

2.2.4 SP_TR_NRDB

Searches in protein sequence databases have become a standard research tool in the life sciences. To produce valuable results, the source databases should be comprehensive, non-redundant, well annotated, and up-to-date. However, the lack of a single protein sequence database satisfying all four criteria has forced users to perform searches across multiple databases. This strategy normally produces complete, but redundant results due to different versions of the same sequence report in different databases.

SP_TR_NRDB (or abbreviated SPTR or SWALL) was created to overcome these limitations (2). SPTR provides a comprehensive, non-redundant and up-to date protein sequence database with a high information content. The components are:

- The weekly updated SWISS-PROT work release. It contains the last SWISS-PROT release as well as the new or updated entries.

- The weekly updated SP-TrEMBL work release. REM-TrEMBL is not included in SP_TR_NRDB, since REM-TrEMBL contains the entries that will not be included into SWISS-PROT, for example, synthetic sequences and pseudogenes.

- TrEMBLnew, the weekly updated new data to be incorporated into TrEMBL at release time.

For genes undergoing alternative splicing, files are provided with additional records from SWISS-PROT and TrEMBL, one for each splice isoform of each protein.

2.2.5 Specialized protein sequence databases

In the following we will give a few examples of specialized protein sequence databases. For a comprehensive list of such and other resources for the molecular biologist have a look at Amos' links on the Expasy server (see *Table 1*).

The Clusters of SWISS-PROT and TrEMBL (CluSTr) proteins (9) database offers an automatic classification of SWISS-PROT and TrEMBL proteins into groups of related proteins based on pairwise comparisons between protein sequences. Analysis using different levels of protein similarity has provided a hierarchical organization of clusters.

The MEROPS database (10) provides a catalogue and structure-based classification of peptidases. An index by name or synonym gives access to a set of 'PepCards' files, each providing information on a single peptidase, including classification and nomenclature, and hypertext links to the relevant entries in other databases. The peptidases are classified into families based on statistically significant similarities between the protein sequences in the 'peptidase unit', the part most directly responsible for activity. Families that are thought to have common evolutionary origins and are known or expected to have similar tertiary folds are grouped into clans. MEROPS also provides sets of files called FamCards and ClanCards describing the individual families and clans. Each FamCard document provides links to other databases for sequence motifs and secondary and tertiary structures, and shows the distribution of the family across the major taxonomic kingdoms.

The G-protein coupled receptors database (GPCRDB (11)) includes alignments, cDNAs, evolutionary trees, mutant data, and 3D models. The main aim of the effort is to build a generic database capable of dealing with highly heterogeneous experimental data pertaining to a specific class of molecules. It is a good example for a specialized database adding value by offering an analytical view of data, which a universal sequence database is unable to provide.

2.3 Proteomics related databases

The term 'proteome' is used to describe the protein equivalent of the genome, for example, the complete set of the proteins encoded by a genome. Databases of two-dimensional gel electrophoresis data like SWISS-2DPAGE (12) and the human keratinocyte 2D-gel protein database from the universities of Aarhus and Ghent have been considered the classical proteomics databases.

However, the rapid rate of genome sequencing leads to an equally rapid increase in predicted protein sequences, most of which have no documented functional role. The challenge is to bridge the gap until functional data has been gathered through experimental research by providing statistical and comparative analysis, structural and other information for these sequences as an essential step towards the integrated analysis of organisms at the gene, transcript, protein, and functional levels.

A number of databases address some aspects of genome or proteome comparisons. The Kyoto Encyclopedia of Genes and Genomes (KEGG) is a knowledge base for systematic analysis of gene functions, linking genomic information with higher order functional information (13). The WIT Project attempts to produce metabolic reconstructions for sequenced genomes. A metabolic reconstruction is described as a model of the metabolism of the organism derived from sequence, biochemical, and phenotypic data (14). KEGG and WIT mainly address regulation and metabolic pathways, although the KEGG scheme is being extended to include a number of non-metabolism-related functions. Clusters of orthologous groups of proteins (COGs) is a phylogenetic classification of proteins encoded in complete genomes (15). COGs group together related proteins with similar but sometimes non-identical functions.

The Proteome Analysis Initiative at EBI has the more general aim of integrating information from a variety of sources that will together facilitate the classification of the proteins in complete proteome sets. These proteome sets are built from the SWISS-PROT and TrEMBL protein sequence databases that provide reliable, well-annotated data as the basis for the analysis. Proteome analysis data is available for all completely sequenced organisms present in SWISS-PROT and TrEMBL, spanning archaea, bacteria, and eukaryotes. The Proteome Analysis Initiative provides a broad view of the proteome data classified according to signatures describing particular sequence motifs or sequence similarities and at the same time affords the option of examining various specific details like structure or searchable functional classification. The InterPro and CluSTr resources have been used to classify the data by sequence similarity. Structural information includes amino acid composition for each of the proteomes, classification by the Homology derived Secondary Structure of Proteins (HSSP) (16), and links to the Protein Data Bank (PDB) (17). A searchable functional classification using Gene Ontology (GO) (18) is also available. The Proteome Analysis Database contains statistical and analytical data for the proteins from completed genomes (19).

2.4 Protein signature databases

Very often the sequence of an unknown protein is too distantly related to any protein of known structure to detect its resemblance by overall sequence alignment, but it can be identified by the occurrence of sequence signatures.

A few databases use different sequence signature recognition methodologies and a varying degree of biological information to characterize protein families, domains, and sites. The oldest of these databases, PROSITE (20), includes extensive documentation on many protein families. Other databases in which proteins are grouped by sequence similarity include PRINTS (21), Pfam (22), BLOCKS (23), and SMART (24).

These sequence signature databases have become vital tools for identifying distant relationships in novel sequences and hence for inferring protein function. Diagnostically, the secondary protein sequence databases have different areas of optimum application owing to the different strengths and weaknesses of their underlying analysis methods (regular expressions, profiles, Hidden Markov Models, and fingerprints).

Sequence cluster databases like ProDom (25), CluSTr (9), SYSTERS (26), and ProtoMap (27) are also commonly used in sequence analysis, for example, to facilitate domain identification. Unlike sequence signature databases, the clustered resources are derived automatically from sequence databases, using various clustering algorithms. This allows them to be relatively comprehensive, because they do not depend on manual crafting and validation of family discriminators; but the biological relevance of clusters can be ambiguous; some may just be artefacts of particular thresholds.

Given these complexities, analysis strategies should endeavour to combine a range of sequence signature databases, as none alone is sufficient. Unfortunately, these databases do not share the same formats and nomenclature, which makes using all of them in an automated way difficult. In response to this Integrated Resource of Protein Families (InterPro), Domains and Functional Sites (28)—has emerged as a new integrated documentation resource for the PROSITE, PRINTS, Pfam, and ProDom database projects. Many other resources have since been added. InterPro allows users access to a wider, complementary range of site and domain recognition methods in a single package.

Release 7.0 of InterPro (July 2003) was built form Pfam 9.0 (5724 domains), PRINTS 36.0 (1800 fingerprints), PROSITE 18.0 (1639 families), ProDom 2002.1 (1021 domains) and several smaller resources. It contained 8547 entries, including families, domains and repeats with over 4,655,000 hits against protein sequences in SWISS-PROT and TrEMBL.

A primary application of InterPro is the computational functional classification of newly determined sequences that lack experimental characterization. InterPro is also a very useful resource for comparative analysis of whole genomes (29) and has already been used for the proteome analysis of a number of completely sequenced organisms.

2.5 Structure databases

The number of known protein structures is increasing very rapidly; these are available through the PDB (17). There is also a database of structures of 'small' molecules, of interest to biologists concerned with protein–ligand interactions, from the Cambridge Crystallographic Data Centre (CCDC).

A number of derived databases enable comparisons of 3-D structures as well as insight into the relationships between sequence, secondary structure elements, and 3-D structure.

Dictionary of secondary structure in proteins (DSSP) (30) contains the derived information on the secondary structure and solvent accessibility for the protein structures stored in PDB. Homology-derived secondary structure of proteins (HSSP) (16) is a database of alignments of the sequences of proteins with known structure with all their close homologues. Families of structurally similar proteins (FSSP) (31) is a database of structural alignments of proteins. It is based on an all-against-all comparison of the structures stored in PDB. Each database entry contains structural alignments of significantly similar proteins but excludes proteins with high sequence similarity, since these are usually structurally very similar.

The Structural classification of proteins (SCOP) (32) database has been created by manual inspection and abetted by a battery of automated methods. This resource aims to provide a detailed and comprehensive description of the structural and evolutionary relationships between all proteins whose structure is known. As such, it provides a broad survey of all known protein folds and detailed information about the close relatives of any particular protein.

Another database, which attempts to classify protein structures, is the CATH database (33), a hierarchical domain classification of protein structures from PDB. There are four major levels in this hierarchy; Class, Architecture, Topology (fold family), and Homologous superfamily.

3 Sequence- and text-based database searches

Depending on the type of data, there are two basic ways of searching the above mentioned databases: using descriptive words to search text databases or using a nucleotide or protein sequence to search a sequence or sequence signature database.

3.1 Text-based database searches

One of the major challenges facing molecular biologists today is working with the information contained in many databases, and cross-referencing this information and providing results in ways, which permit broadening the scope of a query and gaining more in-depth knowledge.

Several developers have identified the need to design database indexing and cross-referencing systems to assist in the process of searching for entries in one database and cross-indexing them to another. The most notable examples of these systems are SRS (34), Entrez (35) and DBGET (36), and Acnuc (37). The Sequence Retrieval System, or SRS, one of the most successful approaches to this problem, is described in some detail in the next section of this chapter.

3.1.1 The SRS approach: a general overview

Started as a Sequence Retrieval System (34), it was originally aimed at facilitating access to biological sequence databases like the EMBL Nucleotide Sequence Database. Today SRS has grown up into a powerful unified interface to over 400 different scientific databases. It provides capabilities to search multiple databases by shared attributes and to query across databases quickly and efficiently. SRS has become an integration system for both data retrieval and data analysis applications. Originally SRS was developed at the EMBL and then later at the EBI. In 1999 LION Bioscience AG acquired it. Since then SRS has undergone a major internal reconstruction and SRS7 was released as a licensed product that is freely available for academics.

The EBI SRS server with its more than 130 biological databases and more than 100 integrated applications and simple analysis tools is a central resource for molecular biology data. All SRS database parsers are available to external users and thus, the EBI SRS server plays an important role as a reference site for most other SRS servers. SRS has gained wide popularity and now there are more than 150 academic installations and many pharmaceutical companies worldwide.

The integrating power of SRS benefits from sharing the definitions of conceptually equal attributes amongst different data sets. That allows multiple-database queries on common attributes, thus adding value since data becomes more valuable in the context of other data. Besides enriching the original data by providing html linking, one of the original features of SRS is the ability to define indexed links between databases. These links reflect equal values of named entry attributes in two databases. It could be a link from an explicitly defined reference in data reference (DR) records in SWISS-PROT or an implicit link from SWISS-PROT to the ENZYME database by a corresponding Enzyme Commission (EC) number in the protein description. Once indexed, the links become bi-directional. They operate on sets of entries, can be weighted, and can be combined with logical operators (AND, OR, and NOT). This is similar to a table of relations in a relational database schema that allows querying of one table with conditions applied to others. The user can search not only the data contained in a particular database but also in any conceptually related databases and then link back to the desired

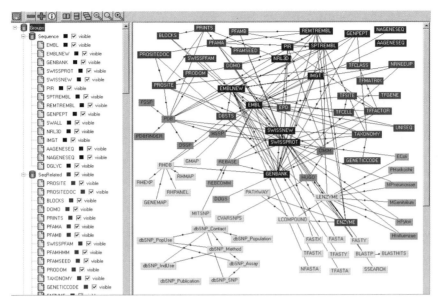

Figure 1 Example of searchable links between databases under SRS.

data. Using the linking graph, SRS makes it possible to link databases that do not contain direct references to each other (*Figure 1*). Highly cross-linked data sets become a kind of domain knowledge base. This helps to perform queries like 'give me all proteins that share InterPro domains with my protein' by linking from SWISS-PROT to InterPro and back to SWISS-PROT, or 'give me all eukaryotic proteins for which the promoter is further characterised' by selecting only entries linked to the Eukaryotic Promoter Database (EPD) from the current set.

3.1.2 The SRS approach: some advanced features

The biosequence object in SRS allows the integration of various sequence analysis tools such as Fasta (38) or CLUSTALW (39); the text output of these applications can then be treated like a text database (see *Figure 2*). Linking to other databases and user-defined data representations becomes possible. More than 100 applications are already integrated into SRS, including tools from the EMBOSS package (http://www.uk.embnet.org/Software/EMBOSS).

SRS also allows the definition of composite views that dynamically link entries from the main query database to other related databases. These views display external data as if they were original database attributes.

An example is the visualization of InterProMatches (InterPro domain composition of a protein sequence) using 'SW_InterProMatches' available at the EBI SRS server. This view dynamically links protein sequences to the InterProMatches database, retrieves information of known InterPro signatures in the proteins and presents the data in a virtually composed graphical form (*Figure 3*).

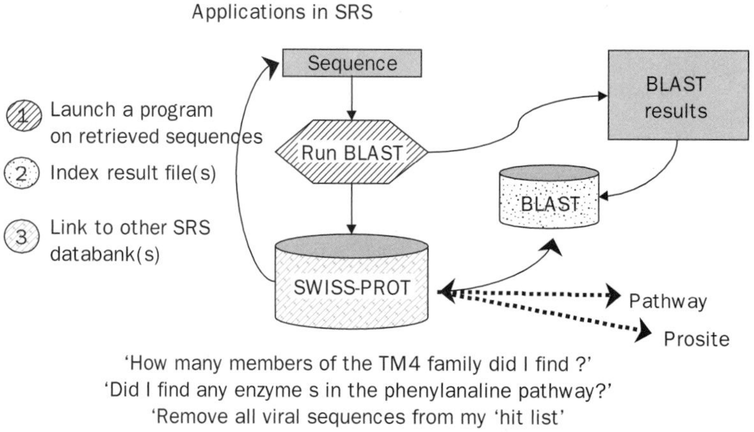

Applications in SRS

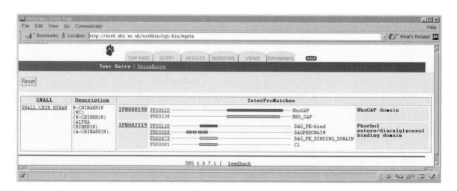

'How many members of the TM4 family did I find ?'
'Did I find any enzyme s in the phenylanaline pathway?'
'Remove all viral sequences from my 'hit list'

Figure 2 The integration of applications in SRS has the advantage of treating the application output like any other database, which allows linking to other databases and user-defined data representation.

Figure 3 Example showing the InterPro domain composition of a protein sequence. The view consists of the ID and Description fields of a protein entry and linked InterProMatches data presented graphically.

3.1.3 NCBI's Entrez

The National Center for Biotechnology Information, a part of the National Library of Medicine, National Institute of Health in the United States, has developed a popular information retrieval system known as Entrez (35).

Like SRS, the Entrez system was specifically developed to deliver sequence and related data across the Internet. Although the number of databases it serves is not as large as some of the major SRS installations (i.e. the EBI's SRS server) it has some data which are unique and extremely important such as PubMed. The other databases currently available through the system include the DDBJ/EMBL/GenBank nucleotide sequence database, SWISS-PROT, PIR, PRF, and PDB for sequence and structural data entry retrieval. Complete genome assemblies, PopSet (a special phylogenetic population study database), Taxonomy and OMIM, as well as many derived datasets such as dbEST, are also available.

Entrez, like SRS, uses an indexing engine to link the various databases available in the system. A notable difference between Entrez and SRS is how the link information is created. While SRS relies wholly on hard-links, Entrez uses the concept of 'neighbours' for describing linked records within a database as well as between records among different databases. Protein and Nucleotide neighbours are determined by searches using the Blast algorithm. Each nucleotide or protein hit is then parsed for literature reference information and cross-linked to the PubMed citation index using the relevant paired model. In this way, a very wide set of neighbour entries is generated which provide reciprocal hard-links as well as closely related information.

Entrez is a completely centralized resource and is only available from the NCBI. To increase the resolution of Entrez search results, the system provides a wide range of options, which are very similar to those available under SRS. These allow specific searches for Subject, Phrases, Authors, Unique identifiers, Molecular weights, and Ranges. Boolean operators (AND, OR, and NOT) are used to combine result sets, and a 'history' facility allows the user to modify previous queries.

3.2 Sequence-based database searches

Only after exhaustive failed searches can it be assumed that a 'new' sequence has not previously been described. But even with a genuinely new sequence, it is possible to find related sequences. Much can be inferred about an uncharacterized protein when significant sequence similarity is detected with a well-studied protein. In the following we will give some basic information on sequence-based database searches.

3.2.1 Pairwise sequence similarity searches: some theoretical background

Alignments provide a powerful way to compare related sequences. Related sequences may be slightly different when comparing a sequence from one organism with a sequence from another, and mutations can give rise to differences between sequences from identical regions in different strains or individuals. Alignments are generally restricted to describing the most common mutations: Insertions, deletions and single-residue substitutions. Insertions and deletions are represented by *null characters* added to one sequence and aligned with letters in the other, while substitutions are represented by the alignment of two different letters.

You can search for nucleotide (or protein) sequences that are just a few bases (or amino acids) in length, or many kilobases (or many amino acids) long. Sequences can be compared by global or by local alignments. Local alignments align only the most similar parts of the input sequences, while a global alignment forces their complete alignment. The choice of a local or a global sequence alignment method depends upon whether the sequences are presumed to be related over their whole lengths or share only regions of similarity.

The huge number of possible alignments is managed by assigning scores. The usual convention for alignment scores is that the higher the score the better the alignment. The most common definition of alignment scores is the sum of scores specified for the aligned pairs of letters and letters with nulls, of which an alignment consists. A *substitution score* is chosen for each pair of letters that can be aligned; the complete set of these scores is called a *substitution matrix*. Advanced statistical matrices have been developed, that allow a query sequence to be aligned with matching sequences in a database. The most significant matches (the best alignments) are reported. The less complex, faster matrices sacrifice a certain degree of match significance that is, you need a better match for it to be recognized than if you use a slower, more complex matrix. The matrix, together with the choice of program essentially determines the search sensitivity and search speed.

A sequence may have a region missing (deletion) or a new region may have been inserted into it as a result of a mutation. When searching a database, it is necessary that the alignment allows for a biologically significant *gap* (inserted or deleted region) in the sequence. This is, for instance, important when aligning a cDNA, essentially a concatenation of exons, against genomic DNA.

For the gaps, which are represented as one or more adjacent nulls in one sequence aligned with letters in the other, scores are chosen. Since a single mutation can insert or delete more than one residue, a long gap should be penalized only slightly higher than a short gap. *Affine gap costs*, which take these considerations into account, are the most widely used gap scoring system. They levy a relative high penalty for the existence of a gap and a smaller penalty for each residue it contains.

The effectiveness of sequence similarity searches depends on the choice of *substitution* and *gap scores*. For *ungapped* local alignments the score for aligning a given pair of residues depends on the fraction (*target frequencies*) of 'true alignment' positions in which these paired residues tend to appear. The desired target frequencies depend upon the degree of evolutionary divergence between the related sequences of interest. It is thus necessary to have a series of matrices tailored to varying degrees of evolutionary divergence to choose from instead of a single matrix. This is the underlying principle in the construction of the Percent Accepted Mutation (PAM) and Block Substitution Matrix (BLOSUM) series of amino acid substitution matrices. Please note that these matrices are generally used unmodified for *gapped* local and global alignments. Since there is no general accepted way of selecting gap costs, their choice needs some bioinformatics intuition.

A set of dynamic programming algorithms is available for finding the optimal (i.e. highest scoring) alignment of two sequences. The first was the *Needleman-Wunsch global alignment algorithm*. The subsequently proposed variant of this algorithm, the *Smith–Waterman algorithm*, can find the optimal local sequence alignment. Both these algorithms require computer processing time proportional to the product of the lengths of the sequences being compared. The most widely used sequence similarity search programmes are based on the Smith–Waterman algorithm, since sequence similarities are often restricted to distinct regions of the sequences compared. However, without specialized hardware the time required by the Smith–Waterman algorithm renders it too slow for most users. The *Fasta* (38) and *Blast* (40) programs use *heuristics* to speed up the alignment procedure. This strategy uses rapid exact-match procedures to identify regions most likely to be related, and only then the Smith–Waterman algorithm is invoked. This approach allows Fasta and Blast to run 10–100 times faster than a Smith–Waterman implementation, at the cost of decreased sensitivity.

The statistics of database searches assume that unrelated sequences will look essentially random with respect to each other. However, many sequences contain regions of highly restricted nucleic acid or amino acid composition and regions of short elements repeated many times. As a result, two sequences containing such compositionally biased regions (so called *low-complexity regions*) can spuriously obtain extremely high match scores.

Most Blast and Fasta servers provide filters, and they are sometimes turned on by default. The filter masks regions of the query sequence that have low compositional complexity by replacing amino acids with a string of X's; a string of N's similarly masks DNA sequences.

3.2.2 Sequence alignment with Blast (basic local alignment search tool)

3.2.2.1 *Overview*

Blast is the algorithm used by a family of five programmes that align a query sequence against the sequences in a nucleotide or protein sequence database. Statistical methods are applied to judge the significance of matches, and alignments are reported in order of significance.

Blastn: Compares a nucleotide query sequence against a nucleotide sequence database.

Blastp: Compares an amino acid query sequence against a protein sequence database.

Blastx: Compares the six-frame conceptual translation products of a nucleotide query sequence against a protein sequence database.

TBlastn: Compares a protein query sequence against a nucleotide sequence database dynamically translated in all six reading frames.

TBlastx: Compares the six-frame translations of a nucleotide query sequence against the six-frame translations of a nucleotide sequence database.

3.2.2.2 *Blast in a nutshell*
Blastp carries out the following six steps during a database search:

1 Identifies all the substitutions of each word of length three that have a similarity score greater than a threshold ($T = 11$). On average, word matches will occur 150 times per database sequence.

2 Builds a hash-like Discrete Finite Automaton (DFA)—a type of pattern-matching program which identifies identical and substituted words.

3 It then uses the DFA to identify all the matching words in database sequences.

4 Once a match is identified it tries to extend it forward and backwards to produce a score that is higher than the threshold score. All scores above the threshold are kept and are reported as the Maximal Segment Pairs (MSPs). These alignments are ungapped.

5 Neighbouring MSPs are combined where possible. For each combination the probability is calculated using either Poisson or sum statistics and the lowest probability scores (P values) are reported. The P value is important. This value expresses the probability, or to say it simply, the number of times you would expect to see such a match by chance. The closer the value is to zero (the smaller the P value is), the less likely the event is.

6 Reports all the significant alignments.

3.2.2.3 *Some blast options*
If you use a Blast server at EBI or NCBI you can tailor your search according to your special requirements by making use of several options. In the following we will explain the two most important options.

The filter option: As explained above, it is very often helpful to use a filter to mask various segments of the query sequence. The SEG program masks low-complexity regions in protein sequences, while XNU masks protein regions containing short-periodicity internal repeats. SEG+XNU will combine the two. The DUST program masks low compositional complexity regions of DNA query sequences. The mentioned programs mask only the query sequence, not the sequences in the database.

The matrix option: Substitution matrices are used both to identify related sequences in a database, and to predict the biological significance of the match. There are two main types of substitution matrices, the PAM and the BLOSUM families of matrices, which are frequently used by sequence similarity search programs.

PAM matrices are most sensitive for alignments of sequences with evolutionary related homologues. The greater the number in the matrix name (e.g. PAM40, PAM120), the greater the expected evolutionary (mutational) distance. You should choose the appropriate matrix for an optimal search. If the mutational distance is unknown, you should run at least three searches using PAM40, PAM120, and PAM250 matrices. You may choose to use PAM to identify conserved sequences or features therein, or to establish the evolution of a sequence.

BLOSUM matrices are most sensitive for local alignment of related sequences. The BLOSUM matrices are therefore ideal when trying to identify an unknown nucleotide sequence. BLOSUM62 is optimized for general protein Blast searches, and is suitable for most situations; it will recognize some amino acid substitutions as conservative (e.g. Arg to Lys). If you are searching for evolutionary related proteins, you should use PAM120 for generalized similarity searches. Be aware that it is not possible to compare the alignment scores from one matrix directly against the alignment scores from another matrix!

3.2.3 Sequence alignment with Fasta

3.2.3.1 *Overview*

The more noticeable difference between Fasta and Blast is the fact that the pairwise alignments in Fasta's output are composed of the best possible global alignment between two sequences. Unlike Blast, Fasta does not report segment pairs between sequences. Although this may seem at first to be a disadvantage because the number of reported alignments (and thus sequence matches) is less in Fasta, it turns out that Fasta reports fewer false positives than Blast. The order in which the scores and alignments are presented in the search report is, as with Blast, according to the Expectation or E( ) value, the probability that a particular match even did not occur by chance.

Below are the main database searching programs available in the Fasta distribution:

Fasta—Compares protein and nucleic acid sequences against protein and nucleic acid sequence databases.

Fastx/y—Compares nucleic acid sequences against a protein database. The programs translate the DNA sequence and compare the translations against a protein database. Fastx uses a simple alignment routine that permits frameshift between codons. Fasty, which is somewhat slower, permits frameshifts to occur within codon and can thus produce better alignments with sequences of poor quality.

TFastx/y—Compares protein sequence fragments against a nucleic acid databases—like tBlastx. These programs are designed to calculate similarities taking frameshifts in the forward and reverse strands into consideration.

TFasta—Compares a protein sequence to a DNA database. Unlike TFastx/y this program calculates similarities without taking frameshifts into consideration.

Fastf/TFastf—This is a recent addition to the package designed to search the protein databases with ordered peptide mixtures obtained by Edman degradation or CNBr cleavage of a protein.

Fasts/TFasts—Another recent addition to the package designed to search protein databases with short peptides fragments obtained via mass spectrometry.

Fasta has many options to control the way it runs. Many have the same or similar functionality as those found for Blast, which were explained earlier. Like Blast, expectation values and expectation value thresholds (max and min) allow the user to fine tune a search and reduce the number of high scoring unrelated sequences once there exists evidence that a biological relationship is being observed. Various options allow choosing between different matrices, gap opening and extension penalty values, etc. Five important options that differentiate Fasta usage from Blast are:

1 Testing using various statistical models. This can thus provide accurate statistics using the '-z' option.
2 The ability to search only sequences within a specific sequence size or size range.
3 The possibility to compare only a fragment of the query sequence during the search. This options works somewhat like a filter and permits the user to concentrate the search only around a fragment of interest in a query sequence.
4 Fasta is much more flexible than Blast for searches that involve very short peptide or nucleic acid sequences.
5 Blast uses a substitution matrix during both the scanning and alignment extension phases while Fasta uses it only during the extension phase.

Points 2 and 3 are not to be confused with Blast's ability to start or stop a search at a particular record number in the Blast database. Another difference worth mentioning is that the Fasta programs can read databases formatted in a variety of different ways. These include GCG, flat files, NBRF, and NCBI's Blast2 format.

The current version of Fasta provides several improvements over older versions. Fasta now calculates optimized scores and provides accurate estimates for statistical significance. The calculation of optimized scores dramatically improves the performance of the programs and although Blast is generally faster, the difference in performance when searching protein libraries is insignificant.

In conclusion, what characteristically differentiates Blast from Fasta is that the latter tries to find patches of regional similarity between a query sequence and the sequences in the database while Blast tries to find the best local alignments. Although Fasta is slower than Blast it compensates with increased sensitivity because it tolerates gaps in the initial alignments.

3.2.3.2 *Fasta in a nutshell*

Fasta performs the following steps during a database search:

1 Identifies regions of identity between two sequences by using a word length of 1 (ktup = 1)—thus being very stringent, or word length of 2 (ktup = 2)—comparing word composed of identical pairs. These regions contain the highest density of identities and once identified are stored for the next step.
2 The 10 regions containing the highest density for step 1 are rescanned using a substitution matrix (default is BLOSUM50). To optimize the score, the ends are

trimmed to include only the region that contribute to the highest score. Each of these regions is then a partial alignment without gaps.

3 If regions from step 2 have a score greater or equal to a threshold or cutoff value they will be further examined to see if they can be joined to form an approximate alignment with gaps. If regions are joined, the similarity score is calculated that is the sum of the score of the individual regions minus penalties for gaps.

4 For all sequences greater than a specified threshold, band optimization is used on the regions reported in step 2 in order to construct an optimal alignment.

5 After the first 60,000 scores the normalized scores or z-values for all sequences which report with ± 5 standard deviations are calculated. Scores above and below these are discarded. The z-values are used to rank the database sequences.

6 The Smith and Waterman algorithm is used to display the final alignment. This alignment has no limitation in the size of gaps and scores low-complexity regions.

3.2.4 Strategies for database searching

The first decision to make when searching molecular databases is which program to use. Blast is the fastest sequence alignment program, but compromises some degree of sensitivity in favour of speed. Fasta is slower, but more sensitive. There is no single best strategy to recommend for searching but the following should serve as a guide.

1 Whenever possible search at the protein rather than at the DNA level using Fasta and/or Blastp. If these searches do not yield a good match or if it is not possible to translate the DNA sequence properly due to errors such as frameshifts, the next step will be to search with the DNA sequence using Fastx/y or Blastx. These programs will translate DNA sequences in all six reading frames and compare each against the protein databases. These searches will quite likely not produce perfect matches overall but will help spot problems with the query sequence. Apart from being very useful for discovering frameshifts they are commonly used to detect contamination which has not been properly removed during contig assembly. Also they help to detect accidental deletions, a most common human error!

2 If results are not conclusive using Fastx/y and/or Blastx the next step is to search DNA versus DNA using Fasta or Blastn, keeping in mind the limitation and merits of each of these programs. It is important to emphasize that the smallest possible DNA database should be searched and only if results are inconclusive to proceed to search larger sets.

3 The very last resort will be to search using TFastx/y or TBlastx using fragments of protein sequences versus the six frame translations of the nucleotide sequence databases. Please note that it is by far the most computationally expensive option and does not guarantee results.

4 It is important to always search the smallest database likely to contain the sequences of interest and only if no biologically significant hits are found to search larger datasets.

5 Use sequence statistics rather than percent identity or percent similarity as your primary criterion for sequence homology and check that the statistics are likely to be accurate by looking for the highest scoring unrelated sequence.

6 The examination of histograms and the extreme value distributions is very important since it allows you to quickly determine if the statistical model used is accurate and helps decide if filtering the sequence or using a different matrix might improve the results. Fasta and Blast plot an asterisk ('*') to represent the expected number of sequences with a given z-score. Equals signs ('=') represent the observed number of sequences. The patterns of '*' in the histogram gives an idea of how well the statistical model used fits the values calculated by the programs.

7 Consider searches with different matrices and different gap penalties. Start with shallow matrices, like PAM100 instead of PAM250, or BLOSUM62 instead of BLOSUM50. Remember to adjust gap penalties to appropriate values when changing matrices.

When assessing results from Fasta or Blast is it important to use sensible E() values thresholds to observe the least possible number of false-positives. There is little one can do about false-negatives and as the database sizes increase so will these. Only biological intuition can help resolve these cases.

4 Final remarks

The databases and text-and sequence-based database search tools are still evolving. However, while the wealth of information in these databases is growing fast there is a lot of molecular biology data still only available in conventional publications. New advances in technology provide even faster means of generating data. It will remain a constant challenge to handle and search the molecular biology data efficiently to facilitate discoveries.

References

1. Stoesser, G. et al. (2003). Nucl. Acids Res., **31**, 17–22.
2. Apweiler, R. (2000). In Advances in protein chemistry, vol. 54 (ed. F. M. Richards, D. S. Eisenberg, and P. S. Kim), p.31, Academic Press, New York.
3. Barker, W. C. et al. (1999). Nucl. Acids Res., **27**, 39.
4. Boeckmann, B. et al. (2003). Nucl. Acids Res., **31**, 365–70.
5. Apweiler, R. et al. (1997). ISMB, **5**, 33.
6. O'Donovan, C., Martin, M. J., Glemet, E., Codani, J.-J., and Apweiler, R. (1999). Bioinformatics, **15**, 258.
7. Fleischmann, W., Möller, S., Gateau, A., and Apweiler, R. (1999). Bioinformatics, **15**, 228.
8. Apweiler, R. (2001). Brief. Bioinfo, **2**, 9.
9. Kriventseva, E. V., Fleischmann, W., Servant, F., and Apweiler, R. (2003). Nucl. Acids Res., **31**, 388–9.

10. Rawlings, N. D., O'Brien, E., and Barrett, A. J. (2002). *Nucl. Acids Res.*, **30**, 343-6.
11. Horn, F. *et al.* (2003). *Nucl. Acids Res.*, **31**, 294-7.
12. Hoogland, C. *et al.* (2000). *Nucl. Acids Res.*, **28**, 286.
13. Kanehisa, M. *et al.* (2002). *Nucl. Acids Res.*, **30**, 42-6.
14. Overbeek, R. *et al.* (2000). *Nucl. Acids Res.*, **28**, 123.
15. Tatusov, R. L. *et al.* (2000). *Nucl. Acids Res.*, **28**, 33.
16. Dodge, C. *et al.* (1998). *Nucl. Acids Res.*, **26**, 313.
17. Berman, H. M. *et al.* (2000). *Nucl. Acids Res.*, **28**, 235-42.
18. Ashburner, M. *et al.* (2000). *Nat. Genet.*, **25**, 25.
19. Apweiler, R. *et al.* (2001). *Nucl. Acids Res.*, **29**, 44.
20. Falquet, L. *et al.* (2002). *Nucl. Acids Res.*, **30**, 253-8.
21. Attwood, T. K. *et al.* (2003). *Nucl. Acids Res.*, **31**, 400-2.
22. Bateman, A. *et al.* (2002). *Nucl. Acids Res.*, **30**, 276-80.
23. Henikoff, S. *et al.* (1999). *Bioinformatics*, **15**, 471.
24. Letunic, I. *et al.* (2002). *Nucl. Acids Res.*, **30**, 242-4.
25. Corpet, F. *et al.* (2000). *Nucl. Acids Res.*, **28**, 267.
26. Krause, A. *et al.* (2002). *Nucl. Acids Res.*, **30**, 299-300.
27. Yona, G. *et al.* (2000). *Nucl. Acids Res.*, **28**, 49.
28. Mulder, N. J. *et al.* (2003). *Nucl. Acids Res.*, **31**, 315-18.
29. Rubin, G. M. *et al.* (2000). *Science*, **287**, 2204.
30. Kabsch, W. and Sander, C. (1983). *Biopolymers*, **22**, 2577.
31. Holm, L. and Sander, C. (1997). *Nucl. Acids Res.*, **25**, 231.
32. Murzin, A. G. *et al.* (1995). *J. Mol. Biol.*, **247**, 536.
33. Pearl, F. M. G. *et al.* (2003). *Nucl. Acids Res.*, **31**, 452-5.
34. Etzold, T. *et al.* (1996). *Methods Enzymol.*, **266**, 114.
35. Schuler, G. *et al.* (1996). *Methods Enzymol.*, **266**, 141.
36. Fujibuchi, W. *et al.* (1998). *Pac. Symp. Biocomput.*, 683.
37. Gouy, M. *et al.* (1985). *Comput. Appl. Biosci.* **1**, 167.
38. Pearson, W. R. (1990). *Methods Enzymol.*, **183**, 63.
39. Thompson, J. D. *et al.* (1994). *Nucl. Acids Res.*, **22**, 4673.
40. Altschul, S. F. *et al.* (1990). *J. Mol. Biol.*, **215**, 403.

Chapter 3

Medical genetics resources in the genomic era

Kim D. Pruitt

National Center for Biotechnology Information, National Institutes of Health, 8600 Rockville Pike, Bethesda, MD 20894, USA.

While the past few decades have yielded impressive gains in our understanding of biology and inherited diseases, the availability of the human genome sequence, polymorphism, and expression data will have a critical impact on the field of medical genetics. The human genome sequence data opens powerful new approaches to integrate and access information. The power, and promise, of medical genetics lies in the ability to intercalate disparate data from multiple sources. At the intersection of genetics, genomics, and medicine, medical genetics holds great promise for understanding the fundamental nature of many diseases, which in turn will lead to better diagnoses, treatments, and even cures.

The numerous web resources relevant to medical genetics are targeted towards the general public, the medical community, and the research community. These sites provide information ranging from patient support groups, to clinical testing laboratories, clinical trials, disease catalogues, epidemiological synopsis, mutation databases, gene family databases, and gene catalogues. Some are broad in scope and present information on a large number of genes or diseases, while others deal with a single disease or a small number of related genes. An entire book could be devoted to this topic to cover this breadth in topics and target audience. This chapter focuses on the impact of the human genome sequence, resources that provide disease and allele oriented descriptions, and some clinical web sites. The resources covered are actively maintained and of medium to large scope.

1 Impact of the human genome project

There have been many articles, editorials, and letters on the anticipated medical applications of the human genome project. The genome sequence is the foundation that both anchors subsequent accumulated data and facilitates a more rapid convergence of genetics, genomics, and medicine. Identifying new genes, polymorphisms, and expression patterns builds upon this foundation. Research over the past couple of decades has greatly expanded our fundamental understanding of the pathogenesis of inborn errors by identifying the genes and products associated with inherited diseases. Several WWW resources, the most notable of which is the Online Mendelian Inheritance in Man (OMIM) (1), catalogue this accrued knowledge. Others catalogue sequence differences including insertions, deletions, and polymorphisms; this information supports efforts to design better diagnostic tests or customized treatment plans. Still other locations provide information on genetic or physical map data. The majority of these resources are scattered across the Internet. Although they do provide cross-links to additional relevant resources, it often takes considerable time and effort to navigate through multiple web sites and cull out information concerning a disease phenotype, function of the causative gene(s), the map position, identified mutations and polymorphisms, and physical clone resources.

The National Center for Biotechnology Information (NCBI) has made a concerted effort to integrate information available at that web site and to provide extensive cross-linking between NCBI resources (2; *Figure 1*). Several complimentary pages at the NCBI encompass an enormous range of information and offer multiple ways to access information about genes, sequence variation, maps, diseases, and clone resources. *Table 1* provides URLs for the NCBI network of human genome resources that are briefly described below. This network also includes OMIM (see section below) and the Mammalian Gene Collection (MGC) (3) project supported by the National Cancer Institute (NCI) of the US National Institutes of Health. Together, these resources facilitate accessing a wealth of information of interest to the medical genetics community. Information is accessible via numerous routes including by starting with a text word query (a gene symbol or name,

Table 1 NCBI network of human genome resources

Resource	URL
Clone Registry	http://www.ncbi.nlm.nih.gov/genome/clone/index.html
NCBI Contig	http://www.ncbi.nlm.nih.gov/genome/guide/build.html
dbSNP	http://www.ncbi.nlm.nih.gov/SNP/index.html
Human BAC resources	http://www.ncbi.nlm.nih.gov/genome/cyto/hbrc.shtml
Human genome BLAST	http://www.ncbi.nlm.nih.gov/genome/seq/page.cgi?F=HsBlast.html&ORG=Hs
LocusLink	http://www.ncbi.nlm.nih.gov/LocusLink/
Map Viewer	http://www.ncbi.nih.gov/mapview/
MGC	http://mgc.nci.nih.gov/
RefSeq	http://www.ncbi.nlm.nih.gov/RefSeq/
UniGene	http://www.ncbi.nlm.nih.gov/entrez/query.fcgi?db=unigene
UniSTS	http://www.ncbi.nlm.nih.gov/entrez/query.fcgi?db=unists

(a)

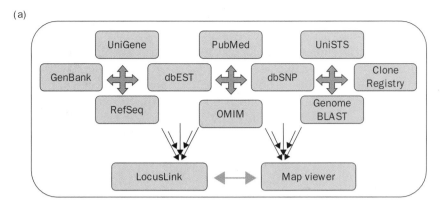

(b)

Start with disease:
- Find disease description (OMIM)
- Navigate to gene information (LocusLink)
- Find associated sequence variations (dbSNP, MapViewer)
- Find tissue expression information (UniGene)
- Find genomic sequence (LocusLink, Map Viewer)
- Find other genes in the genomic region (Map Viewer)
- Find available BAC or cDNA (Clone Registry, MGC)

Start with position:
- Find position on map (Map Viewer)
- Determine if the genomic sequence is available (Map Viewer)
- Find genes in that region (Map Viewer)
- Find gene information (LocusLink)
- Find associated sequence variations (dbSNP, MapViewer)
- Find tissue expression information (UniGene)
- Find available BAC or cDNA (Clone Registry, MGC)

Figure 1 The NCBI Network of human genome resources. (a) Extensive cross-linking is provided between NCBI resources making it possible to begin with disparate types of information and easily navigate to information available in other resources. The primary human genome resources are indicated here. LocusLink serves as a central hub and provides direct links to numerous resources. The Map Viewer also provides several links; however, the links provided depends on which maps are included in the display. Additional links can be accessed by modifying the display configuration. (b) Two examples illustrate some of the information that could be accumulated when starting with different types of information. The resource providing distinct information types is indicated in parenthesis. The order of the list does not represent a direct navigation path; some information is obtained by following multiple paths from LocusLink. The provision of multiple access routes to the same network of information facilitates information retrieval and expands the available discovery space.

a sequence tagged site (STS) marker name, a BAC clone name, a disease name, a sequence accession number), a sequence (using BLAST), or a position on a chromosome. As illustrated in *Figure 1*, it is possible to rapidly find a considerable amount of information even when beginning the search using very different types of information, such as with a disease name, a cytogenetic position, a STS marker name, or a physical sequence. Each resource that contributes towards the overall network of human genome data is briefly described below; additional information about each is available on-line.

1.1 dbSNP

The Database of Single Nucleotide Polymorphisms (dbSNP), a variation database, includes sequence information about single nucleotide polymorphisms, insertions, and deletions. The Map Viewer display shows SNP density along the chromosome and indicates coding region SNPs.

1.2 Human BAC Resource and Clone Registry

The Human BAC Resource site provides a genome-wide resource of large-insert clones (bacterial artificial chromosomes) for which both cytogenetic Fluorescent *in-situ* hybridization (FISH) map and sequence information is available; these clones help integrate cytogenetic, radiation-hybrid, linkage, and sequence maps. The Clone Registry records the sequencing status and distributor information for BAC clones. It is possible to start at either of these pages with a clone name and retrieve information about the clone including reagent availability and navigation to the Map Viewer resource where additional information is also available.

1.3 Human genome BLAST

A custom BLAST (4) page accommodates carrying out a sequence similarity search against the genomic contigs, the NCBI reference sequences (RefSeq) transcripts (including the models annotated on the contigs), and the RefSeq proteins. This resource then provides the means to start with a sequence of interest and determine if there is genomic sequence that corresponds to it. Furthermore, it is possible to navigate to the Map Viewer from the genome BLAST result page (click on the Genome View button). The BLAST result is displayed in the context of the genome—thus providing rich information about genomic structure (if query was a cDNA) and what other genes, STS markers, SNPs, and clone reagents are available in that location.

1.4 LocusLink

LocusLink is a gene-centred resource that provides information on nomenclature, publications, gene ontology terms (GO), computed protein domains, map position, sequence data (both GenBank and RefSeq) (5, 6). LocusLink reports indicate when additional information is available in related resources such as the Map Viewer, OMIM, dbSNP, UniGene, PubMed, and much more.

1.5 Mammalian gene collection

The MGC project generates full-length complementary DNA (cDNA) sequences and clone resources (3). Search for available cDNA reagents by starting with a MGC accession number or gene symbol, or by submitting a sequence of interest to a BLAST query.

1.6 NCBI Map Viewer

The Map Viewer provides a robust query interface, a genome-scale overview of query results, and configurable chromosome-level tabular and graphical views of numerous maps including radiation hybrid (RH), genetic, cytogenetic, and the human genome sequence data. NCBI is assembling the human genome sequence into larger contiguous sequences—contigs—and applying annotation to those contigs. The annotated contigs are displayed in the Map Viewer and facilitate public access of the genome sequence data by providing a non-redundant genome assembly that incorporates biological context information (genes, varia-tion, markers, etc). The Map Viewer displays the order and location of the contigs, the corresponding GenBank sequence data used to assemble the contigs, the location of genes, STS markers, SNPs, and clones as well as other map data.

1.7 RefSeq

The NCBI RefSeq project provides curated RefSeq(s) records for numerous organ-isms including human (6). The human RefSeq records represent a non-redundant sequence database that uses current nomenclature and uniformly incorporates additional information (as known) including alternate names, EC numbers, map location, publications, and computed protein domain information. In addition, corrections and extensions to the sequence are identified through an on-going review process of the available sequence and literature data; this review also incorporates additional biological information in the form of brief summary descriptions of the gene function and additional feature annotation on the nucleotide and protein records.

1.8 UniGene

UniGene is an experimental system for automatically partitioning GenBank sequences into a non-redundant set of gene-oriented clusters with related information such as the tissue types in which the gene has been expressed and map location (7). In addition to sequences of well-characterized genes, hundreds of thousands of novel expressed sequence tag (EST) sequences have been included. Consequently, the collection is useful to the community as a resource for gene dis-covery. UniGene has also been used to select reagents for gene mapping projects and large-scale expression analysis.

1.9 UniSTS

UniSTS reports information about markers, or STS, integrated with mapping data from public resources. Information displayed includes the primer sequences, product size, mapping information, and cross references to other resources including the Map Viewer. The NCBI electronic PCR (e-PCR) tool uses UniSTS data to identify STS markers that match a query sequence and in effect electronically map a query sequence to a chromosome location based on matches to mapped STS markers.

2 Disease and gene resources

The disease and gene resources described below are the predominant, larger-scope resources available on the WWW (see *Table 2*). These catalogues, or encyclopedias, of diseases and genes, serve an important function in summariz-ing our current state of knowledge. They make a large volume of historical and current literature information conveniently and easily accessible and thereby represent an essential resource to the research community. The database and

Table 2 Disease- and gene-oriented resources

Resource	URL
GeneCards	`http://bioinformatics.weizmann.ac.il/cards`
GeneTests	`http://www.genetests.org`
Genes and Disease	`http://www.ncbi.nlm.nih.gov/books/bv.fcgi?call=bv.View..ShowTOC&rid=gnd.TOC&depth=2`
HuGENet	`http://www.cdc.gov/genomics/hugenet/`
LocusLink	`http://www.ncbi.nlm.nih.gov/LocusLink/`
OMIM	`http://www.ncbi.nlm.nih.gov/entrez/query.fcgi?db=OMIM`

WWW availability of these resources also makes it possible to provide valuable connections to other sources of information—thus the basic process of carrying out preliminary literature research to survey what information is known is hugely supported by these online resources. Without them it would be a very laborious process indeed to compile the disease and gene information, plus the knowledge of additional genomic or clinical sources of information, from the literature.

2.1 Online Mendelian Inheritance in Man™

OMIM (1) provides a continually updated catalogue of inherited genetic disorders and genes and is a significant medical genetics web resource for the clinical, medical, and research communities. With over 14,600 entries OMIM, considered to be the phenotypic companion to the human genome project, is the most comprehensive catalogue of this type currently available. OMIM is authored and edited by Dr Victor A. McKusick and his colleagues at the Johns Hopkins University, and the web site is maintained by the NCBI. It provides disease and gene report pages plus separate reports of the cytogenetically localized genes and diseases (the OMIM Gene Map) and all disorders included in OMIM (the OMIM Morbid Map). This site also provides supporting Help and FAQ pages.

This site has a powerful query interface, as the OMIM database is included in NCBI's *Entrez* query and retrieval system. The interface supports flexible field searching and provides a range of capabilities including a 'Limits' function that easily allows adding search restrictions such as chromosome number and other record attributes. It is possible to select and display multiple records in a number of formats. In addition, the automatic cross-linking of all *Entrez* databases means that OMIM pages are automatically linked to related *Entrez* records including PubMed and GenBank. This cross-linking works in both directions; thus, upon querying PubMed or the GenBank nucleotide database, the resulting page includes links back to OMIM when the retrieved item (for example, a publication) is included in OMIM.

Most pages within the site include a side column with navigation links to related pages or to content within the page displayed. The more general pages such as the home page and search page include a standard set of links to information including the OMIM Gene and Morbid maps, help documentation, a FAQ page, statistics, and other useful web sites. OMIM disease and gene review pages provide links to sections within the page.

2.1.1 Query

The query interface supports wildcard (*) and Boolean (AND, OR, NOT) queries that can be further refined by using the 'Limits' and 'Preview/Index' settings. The History page stores all queries done during a session; this provides a very convenient way to return to a previous query without having to redefine the query itself. In addition, query results can be temporarily saved to the Clipboard where items can then be displayed or saved. This service is a useful way to collect a subset of results from a large result set, or to collect records of interest that must be found using multiple distinct queries. The extensive OMIM Help document provides details and examples of basic and advanced queries; the main features are highlighted below.

2.1.1.1 *Query restrictions*

The OMIM Limits setting permits one to restrict a query by field, chromosome, type of record, or by inclusion of specific subsections. More than one item can be selected for each category. For example, selecting one or more items from the Fields category will restrict the query to identify only those results with matches to the selected field(s) such as title, MIM number, allelic variants, text, references, clinical synopsis, gene map disorder, and contributors. Likewise, it is possible to restrict a query to a type of MIM record such as those for which the mode of inheritance is known, or records in which a single phenotype may be caused by different mutations or genes. Queries can also be restricted to records that include specific information such as allelic variant or clinical synopsis sections. Query restrictions can be selected in the Limits form; more than one item can be selected from each of the four areas. For example, a query for ataxia with no Limits set returns over 500 results. Using the Limits settings to restrict matches to the title field, to records for which the mode of inheritance is proved, and to records that include a clinical synopsis section reduces the results to a more manageable 41 records.

The Preview/Index form serves multiple functions. It provides a term list which can be used to further refine a query, a preview function which indicates the number of results found, and the opportunity to browse all of the indexed words (e.g. those words to which query words are compared in order to find a result). Previewing is useful for fine-tuning an initial query that returned an unwieldy result set. For example, use the preview function with each successive change to the query to determine if the results set is winnowed down to a manageable size. To preview the actual indexed words simply select a term from the pull down menu (such as Allelic Variant, Clinical Synopsis, or Properties), enter a term in the text box (e.g. bind*) and click on the 'Index' button. This presents a list of phrases and words and their frequency of occurrence. You can select a word or phrase from the menu to add to your query by clicking on the word of interest and on one of the 'AND, OR, NOT' buttons. For example, if interested in the intersection of DNA repair as it relates to cancer you can refine an initial query for 'cancer' by selecting Clinical Synopsis from the term menu, entering 'DNA' into the text field, and clicking on the 'Index' button. Select 'dna repair(9)' from the menu list, and click on the 'AND' button to add that to your query. Previewing the query results shows there are eight results, and clicking on the 'GO' button will retrieve those results.

2.1.1.2 *Navigating Entrez query results*

Upon refining a query, the NCBI *Entrez* system provides many useful cross-links. A document summary (docsum) page displays the query results. Items of interest can be selected individually to view the OMIM report (click on the MIM number), or several results can be selected by clicking the check box. Select one of the display options from the Display menu list to modify the format of the docsum page. For example, given a query that returns 2 pages of results, it is possible to select a small number of records by checking the box to the left of each result. The display can then be changed to show a number of different fields incorporated into many OMIM records. For example, you can elect to display the clinical synopsis, allelic variants, sequence links, PubMed, or LinkOut links for the set of selected results.

2.1.2 Disease and gene report pages

The disease and gene report pages are the key element of the OMIM web site; they present detailed descriptions of diseases and genes and provide the critical connections to the sequence data. These entries provide a heavily cited historical overview organized by topic. The extensive bibliography alone is extremely valuable, and when coupled with text that provides brief descriptions of the main conclusions of each reference, these articles represent a significant resource for the medical genetics community. New entries continue to be added and pre-existing entries are updated, as new information becomes available. Reports include an edit history so it is possible to see the date at which the record was last updated. Some of the older, longer, entries can be difficult to read as the text includes the nomenclature used in the article being referred to, and so a single entry may refer to alternate protein or disease names used in the literature. Nevertheless, entries provide a valuable summary of the available literature and include descriptions of several topics (as relevant) including:

- allelic variants
- animal models
- biochemical features
- clinical features
- clinical management
- diagnosis
- gene cloning

- gene function
- inheritance information
- mapping
- molecular and population genetics
- references.

Navigation within a report is conveniently supported by a series of links in the left column; the links are always visible on the page so as you scroll down so it is easy to jump to a different section. Links are also provided to the Mini-MIM and Clinical Synopsis when available. The Mini-MIM is a condensed version of the primary report and is useful for getting a quick summary. The full report includes additional details and may also be more current. The clinical synopsis page provides a good overview of the clinical features in a simple list format. Additional links are provided that greatly facilitate navigation to and between related NCBI resources including LocusLink, GenBank, Map Viewer, PubMed, dbSNP, as well as to external resources such as the Coriell Cell Repository and Mutation databases. This wider network of links provides a powerful tool to the research community by facilitating access to a considerable amount of related human genome information.

2.1.3 Gene Map

The OMIM Gene Map presents the cytogenetic map location of the genes described in OMIM. Not all OMIM entries are represented in the Gene Map, but only those for which a cytogenetic location has been published. The OMIM Gene Map is available on the web and FTP site. The web version of the OMIM Gene Map can be searched with a gene symbol, chromosomal location, or disorder keyword. Results are returned for all matches between a query term and any of the Gene Map fields.

The result page displays the first location that matches the provided query term. The 'Find Next' button can be used to find subsequent instances of the query term. For example, a query for 'hemoglobin' returns a match to 'Hereditary persistence of fetal hemoglobin' (MIM: 142470) on chromosome 6. Following the 'Find Next' link returns a page starting with the chromosome 7 HPFA2 gene (MIM: 142335). Gene Map results are displayed 20 entries at a time and 'Move up' and 'Move down' links allow you to page up and down the map. The tabular result page indicates the region displayed as both the cytogenetic range and the genes defining the top and bottom boundary of the view. Columns present the cytogenetic location, gene symbol(s), OMIM title, MIM number, disorder, comments, and mouse homologue. The cytogenetic location is linked to NCBIs Map Viewer where the OMIM Gene Map data is included in the 'Genes_cytogenetic' map (additional sources of cytogenetic data also contribute to the map display). The MIM number is linked to OMIM disease and gene report pages, and the mouse homology links are directed to gene reports maintained by The Jackson Laboratory.

2.1.4 Morbid Map

The OMIM Morbid Map is an alphabetical list of diseases described in OMIM and their corresponding cytogenetic locations (8). It is available on the web and the FTP site. The web version can be searched with a gene symbol, chromosomal location, or disorder keyword. The result page displays 20 diseases at a time and starts at the first match to the provided query term. The 'Find Next' button takes you to subsequent instances of the query term, or you can scroll up and down 20 entries at a time using the 'Move Up' and 'Move Down' links.

The query result page lists the disorder, gene symbol, MIM number, and cytogenetic position (if known) in tabular format. When there are separate disease and gene report pages, links to both reports are provided in the Disorder and OMIM columns, respectively. The Morbid Map is also available as a graphical display in the NCBI Map Viewer. This graphical display shows the OMIM disease genes in cytogenetic position. Use the Map Viewer 'Display Settings' form and select the 'Morbid' map to display.

2.2 Genes and Disease

The NCBI Genes and Disease resource presents short descriptions of diseases and their associated genes for students and the general public. The reviews provide a good general description of the diseases covered and focus on diseases that have been highlighted in the literature recently. Reviews are provided as an on-line book in the NCBI Books database and can be accessed from a table of contents. New chapters and reviews continue to be added. Topics range from cancer to immunology to organ systems and development. Several reviews are available under each chapter; reviews discuss the disease, associated genes, and provide useful links to related resources hosted at NCBI (such as the Map Viewer, PubMed, RefSeq LocusLink, and OMIM) as well as links to other NIH institutes, support groups, and other relevant resources.

2.3 GeneReviews

GeneReviews presents peer-reviewed clinical genetic information authored by experts in the field and therefore provides a significant medical genetics resource (9). GeneReviews, available from the GeneTests website, provide a significant source of medical, genetic, and clinical information for those diseases currently included. Although its scope is more limited than OMIM, GeneReviews currently includes very thorough descriptions of over 200 disorders for which clinical tests are available. Each disease page provides a synthesis of current understanding of the disorder with a focus on the use of genetic testing for diagnosis, treatment management, and patient counseling. The original date posted, as well as the last update date, are available at the top of each report, making it easy to determine how current the information is. The web site is specifically targeted for use by genetic counselors and medical geneticists.

Follow the GeneReviews link at the top of the home page to access disease reports via a query or by browsing the list of titles (follow the 'Title' link in the side column). When carrying out a query, all matches to the query term are listed and a link is provided to the corresponding GeneReview page when one is available. The query result is often presented in a hierarchy with related diseases grouped together.

Report pages are well organized and include navigational links to different topics covered; each article provides a general overview of each disorder as well as the diagnostic approach and clinical description of the disorder that are suitable for use in diagnosis. The disease articles include abundant references with links to PubMed. Available clinical tests are detailed and address the complexities of differential diagnosis is due to non-specific clinical presentation. Known gene mutations and frequency of occurrence are described, however there are no direct links to corresponding DNA or protein sequence data. Entries also include information on genetic risk assessment and the use of family history and genetic testing to provide family counseling. Reports do include links to several related resources including OMIM, GDB, LocusLink, GeneCards, GenAtlas, the Human Gene Mutation Database (HGMD), and to other locus-specific mutation resources.

2.4 Human Genome Epidemiology Network (HuGENet)

Epidemiologic studies assess and quantify the impact of allelic variation on disease, and identify population and environmental factors that interact with gene variants to alter disease risk, mortality, or treatment. These studies also provide a baseline to validate the effectiveness of new clinical tests for different population groups. HuGENet is an international collaborative effort to disseminate population-based epidemiologic data relevant to the human genome (10). This resource is provided by the US Centers for Disease Control (CDC) and information is provided as a series of articles organized by specific genome–disease interrelationships. The HuGENet site is well organized and includes intuitive navigation links to background information, the reviews, and related information at the CDC. HuGENet does not provide a query interface to search the reviews; available reviews are simply presented in a list format that also indicates review topics currently under development. Articles are periodically updated and the update date is displayed so it is quite easy to determine how current any given article is.

Articles are written by experts on the subject and are peer-reviewed by an editorial board prior to publication. Although there are a relatively small number of articles currently available, each one provides an in-depth comprehensive overview of known allelic variants and their frequency of occurrence in different populations. Each review describes the gene, protein function, and the associated disease including phenotype, clinical symptoms, treatment options and efficacy, complications, and prognosis. These reviews also provide information on the magnitude of risk, contradictory reports, other associated factors, and the effectiveness of genetic tests. Population frequency information is provided as a summary of the literature. Defined queries of the PubMed database are used to identify papers to consider; studies of small sample size are not included in the frequency survey. The literature survey is used to identify both genotype frequency and geographic location.

These reviews include a rich bibliography citing papers describing the gene, disease, allelic variants, and population studies. While they include useful descriptions of sequence changes associated with disease, reviews do not explicitly establish what the baseline of comparison, or RefSeq, is; there is no direct connection to sequence data. Allelic variants are most meaningful in the context of a RefSeq so it would be useful if reviews included accession numbers. Since the sequence connection is indirectly accessible through the references, it is possible to navigate from a publication in PubMed to find those sequences that include that citation.

2.5 LocusLink

LocusLink provides a single query interface to curated sequence and descriptive information about genetic loci (5). It presents information on official nomenclature, aliases, sequence accessions, computed protein domains, phenotypes, EC numbers, publications, GO terms, MIM numbers, UniGene clusters, homology, map locations, and related web sites. The query engine supports the use of wild cards (*), Booleans (AND, OR, NOT), and some field restricted queries (e.g. symbol, or chromosome). LocusLink reports serve as a central-point-of access for information about human genes; reports summarize available information and provide links to related resources where additional information is available. One notable distinction of the LocusLink resource is that data is subject to an ongoing review to increase the accuracy and comprehensive quality of the data. Thus LocusLink reports may include alternate names used over time for the same gene, and some reports include a concise summary of the gene function. LocusLink also provides a mechanism for the community to submit additional functional information as a concise statement of the functional significance attributed to the gene in a published report; all gene report pages include a link to the LocusLink 'Gene References into Function' (GeneRIF) submission form.

2.6 GeneCards

GeneCards is a database of human genes provided by the Weizman Institute of Science (11). This resource provides a single point of access to data that is downloaded from other resources. Gene report pages provide an integrated view to the accumulated information and thus provide a convenient way to access an overview of available information about a gene. The information accrued includes nomenclature, map position, disease associations, and sequence associations. However, since data is obtained by downloading other resources, errors extant in those resources will be propagated to GeneCards entries, as additional curation work is not provided. The web site includes a query interface and documentation on the sources of information and process used to integrate it into GeneCards reports. Reports are concise and indicate the original source of information.

3 Mutation and polymorphism databases

One of the goals of the Human Genome Project is to assemble a catalogue of common human sequence polymorphisms (12). Variation data, in the form of insertions, deletions, and SNPs, provides data that is essential to current and future efforts to map complex disease traits using whole-genome association studies (12). The two major SNP archival databases, NCBI's dbSNP (13) and the European Human Genic Bi-Allelic Sequences Database (HGBase) (14) provide catalogues of human sequence variation that facilitate research into how genotypes affect common diseases, drug responses, and other complex phenotypes. Knowledge of known mutations and biologically relevant polymorphisms plays a critical role in the diagnosis of many genetic diseases and in locating the best tissue donor matches. As additional mutation and polymorphism data accumulate, it is expected to play a significant role in the development of more sensitive diagnostic tests and customized disease treatment options.

Although gene- and disease-centred resources provide significant information about phenotypes, clinical descriptions, diagnosis, and treatment, their connection to the underlying mutation or polymorphism sequence data is not always as robust as it could be. The most useful polymorphism and mutation databases are those that provide a direct connection to a sequence standard in addition to compiling an index of the known sequence changes. These resources range from the large HGMD (with over 1300 genes), to the mitochondrial database MITOMAP (15), to gene cluster or family databases, to single gene mutation databases. Furthermore, several resources provide listings of mutation database sites. It is not possible to review all of the individual mutation databases; it is challenging to even compile a complete listing of them. The large public polymorphism databases have been described in detail elsewhere (13, 14); other allele and mutation oriented resources are described below. *Table 3* lists additional URLs for variation and mutation databases as well as some resources that provide links to several mutation web sites. One of these, the GenomeWeb, focuses on collecting links to a range of genomic, map, sequence, and genetic web resources and includes a large list of mutation databases.

3.1 The Human Gene Mutation Database

The HGMD provides a significant reference source for known mutations underlying human genetic disease (16); the mutation catalogue is limited to those occurring in the coding region. The web site integrates phenotypic, literature, and sequence data to provide information of practical diagnostic importance to researchers in human molecular genetics, physicians interested in a particular inherited condition, and genetic counselors. New mutation information is added

Table 3 Polymorphism and mutation databases

Resource	Type[a]	URL
dbSNP	V	http://www.ncbi.nlm.nih.gov/SNP/index.html
GenomeWeb	L	http://www.hgmp.mrc.ac.uk/GenomeWeb/ human-gen-db-mutation.html
HGBase	V	http://hgvbase.cgb.ki.se/
HGMD	M	http://archive.uwcm.ac.uk/uwcm/mg/ hgmd0.html
HGMD	L	http://archive.uwcm.ac.uk/uwcm/mg/docs/ 0th_mut.html
HUGO mutation database	L	http://www.genomic.unimelb.edu.au/ mdi/dblist/glsdb.html
IMGT/HLA	M	http://www.ebi.ac.uk/imgt/hla/
MITOMAP	M	http://www.mitomap.org/
Mutation resources list (MutRes)	L	http://srs.ebi.ac.uk/srs6bin/cgi-bin/ wgetz?-page+LibInfo+-id+4Flds1FDMZh+-lib+MUTRES
OMIM	L	http://www.ncbi.nlm.nih.gov:80/entrez/Omim/ allresources.html#LocusSpecific
Universal mutation database	L	http://www.umd.necker.fr/disease.html

[a] L, links to mutation databases; V, archival variation database; M, mutation or allele database.

at a regular basis and is available on the web site following an undefined period of exclusive access by the private company Celera. Unfortunately there is no indication of the length of the exclusive access period at this time. Nevertheless, the resource provides a convenient integrated summary of mutations and provides the essential direct connection to a sequence baseline. The resource provides a convenient query interface, and is useful way to identify known mutations.

Information is found by searching with a gene symbol, gene name, accession number, OMIM ID, or GDB ID. Report pages identify each gene by name and symbol. A short bibliography is provided, followed by a table listing the number of described mutations as organized by type (e.g. number of substitutions, deletions, insertions, etc). A separate page listing each mutation is provided for each category (category names are linked to these pages); mutations are described by codon position, nucleotide change, amino acid change, disease association, and the publication that first described it. The main report page also lists diseases associated with the gene, with links to OMIM, and the number of known mutations associated with each disease.

Towards the bottom of the HGMD Report page smaller tables provide links to associated data, HGMD navigation links, and links to external resources. The Associated data provides a graphical Mutation map that marks the relative location of each mutation on the protein; this provides a useful overview that serves to highlight regions with a higher number of mutations. The Associated data section also provides a link to a reference cDNA sequence page where the coding region sequence and GenBank accession number are available. This direct association between a reference cDNA sequence and associated mutations is essential to the research group interested in using this information. Without an explicit reference point, it can be difficult to navigate from a reported mutation to the precise location on a representative sequence, since there are often many representative sequences available in the public sequence databases, and differences between these sequences might be attributable to polymorphism or sequence error. The primary limitation is that this mutation database collects only coding region mutations and thus omits mutations that occur in introns and regulatory regions. Report pages include links to external sites such as OMIM, GDB, GenAtlas, and the Human Gene Nomenclature Committee.

3.2 Immunogenetics Human Leukocyte-Associated Antigen Database (IMGT/HLA)

The IMGT/HLA database (17) provides an critical resource to the tissue typing community by maintaining a database of nucleotide and protein sequences of the polymorphic genes of the HLA system, the human major histocompatibility complex (MHC). This database includes sequence information for the highly polymorphic HLA class I and class II alleles that influence the success of organ transplants, as well as for genes associated with disease susceptibility or the progression of some infectious diseases. The IMGT/HLA group also has an integral role in assigning standardized nomenclature for newly identified alleles. The sequence entries currently available include cDNA and protein data; there are plans to incorporate intron sequence data in the future. This resource offers the advantage that all sequences are reviewed and annotated by an expert before acceptance into the database; thus, known or suspected sequencing errors can be avoided and sequences are associated with the correct allele name.

Information can be accessed multiple ways including querying for an allele or cell line, browsing a preformatted alignment, or performing a sequence similarity search. The database is also integrated into the European Bioinformatics Institute (EBI) SRS retrieval system. The database is updated in periodic releases and in addition to the facilities mentioned above, it provides an interface to specify a new alignment view, to submit new alleles, and includes links to other relevant resources. A companion cell line database contains descriptions of the cell line from which each allele was isolated; the cell database can be used to find alleles by cell line, DNA type, serological type, or ethnic origin.

4 Clinical resources

As more is learned about human genetics, physicians, clinicians, researchers, and the lay public increasingly utilize information about the availability of genetic testing for specific disorders, how genetic testing is used, and the accessibility of new clinical trials. Internet resources that provide information on diagnosis, clinical testing laboratories, and clinical trials vary widely in their scope and target audience. Some representative resources that are more comprehensive and are targeted towards the clinical or research community are discussed below.

Table 4 Clinical resources

Resource	Type[a]	URL
Cancer Clinical Trials	T	http://cancertrials.nci.nih.gov/system/
ClinicalTrials	T	http://clinicaltrials.gov/
GeneTests	T	http://www.genetests.org/
Genetic Alliance	S	http://www.geneticalliance.org/
Information for Genetics Professionals	L, S	http://www.kumc.edu/gec/geneinfo.html
OncoLink	L	http://oncolink.upenn.edu/

[a] L, links to clinical resources; S, information on support groups; T, information on clinical trials.

Table 4 lists some additional web sites, such as the University of Pennsylvania's OncoLink (18), which provide useful links to related web site.

4.1 GeneTests

GeneTests provides genetic counseling and testing information for families and healthcare providers; GeneReviews and GeneTests are complementary resources that are extensively cross-linked (9). Information available includes a directory of US laboratories providing clinical testing for inherited disorders. Limited information on international clinical labs is also provided; navigate to the 'Clinical Directory' page and then follow the link to 'Other Clinical Resources'.

Registered users can use this web resource to identify clinical labs or access general information about genetic testing and services including a power point slide presentation that highlights the use of GeneTests in locating and providing genetic services. Clinical labs can be found by querying for a disease of interest, or by looking for labs by location. Clinics listed in GeneTests focus on genetics and prenatal diagnosis of hereditary disorders. This site also includes information pertaining to related services such as DNA banks and genotyping labs.

4.2 ClinicalTrials

ClinicalTrials provides easy access, for patients and health care professionals, to information on international clinical trials for a wide range of diseases and conditions. The web site includes organized general information such as FAQ pages, definitions, and search instructions that clearly outline how to use the resource. This resource provides descriptions of clinical trials summarizing the purpose, disease, and therapy being studied, recruitment status, patient participation criteria, location, and contact information. Additional links to relevant online health resources are also provided.

4.3 Cancer Clinical Trials

The NCI web site provides a wealth of information on types of cancer, testing and diagnostics, genetics, statistics, and clinical trials. Information is provided for the general public, health care professionals, and the research community. There are two main branch points from the NCI home page to access scientific information. The 'Cancer Information' page provides access to descriptions of types of cancer, testing and treatment, and a database of clinical trials. The 'Research Programs' page includes links to general information documents and to resources including the Cancer Gene Anatomy Project (CGAP) pages (19). The CGAP pages provide access to cDNA library expression profiles (follow the 'Tissues' link), cancer-related genes and variation data (the 'Genes' link), chromosome aberrations (follow the 'Chromosomes' link), clone reagents, and analytical tools.

The CancerTrials pages include access to a database of cancer-related clinical trials. Clinical Trial data is stored in the Physician Data Query (PDQ) database. A flexible interface facilitates querying the PDQ by type of cancer, type of clinical trial, international location, and more. Results can be displayed in two configurations targeted towards patient or health professional use. The health professionals report pages include a description of the objectives, protocol, principle investigators, and other relevant information.

5 Conclusions

Although the WWW represents a rich information source for the medical genetics community, it can be extremely time consuming to identify relevant resources. Yet, clearly these resources provide critical information to the medical genetics community, as evidenced by the increased number of articles and web sites that provide directories of available resources (such as the GenomeWeb site, see *Table 3*). Additionally, web sites are increasingly aware of the need to provide cross-links out to other disparate but related resources. These networked resources provide an essential support to the research effort at multiple levels. They make general background information more accessible and they facilitate navigation to other sources of information. It is this opportunity—to conveniently travel to other related sources of information—that leads the way to new discoveries and greater integration of the genetic, genomic, and medical fields.

References

1. McKusick, V. A. (1998). *Mendelian inheritance in man. Catalogs of human genes and genetic disorders*, 12th edn. Johns Hopkins University Press, Baltimore.
2. Wheeler, D. L., Church, D. M., Federhen, S., Lash, A. E., Madden, T. L., Pontius, J. U., Schuler, G. D., Schriml, L. M., Sequeira, E., Tatusova, T. A., and Wagner, L. (2003). Database resources of the National Center for Biotechnology. *Nucl. Acids Res.*, **31**(1), 28–33.
3. Riggins, G. J. and Strausberg, R. L. (2001). Genome and genetic resources from the Cancer Genome Anatomy Project. *Hum. Mol. Genet.*, **10**(7), 663–7.
4. Altschul, S. F., Gish, W., Miller, W., Myers, E. W., and Lipman, D. J. (1990). Basic local alignment search tool. *J. Mol. Biol.*, **215**, 403–10.
5. The Gene Ontology Consortium (2000). Gene Ontology: tool for the unification of biology. *Nature Genet.* **25**, 25–9.
6. Pruitt, K. D., Tatusova, T., and Maglott, D. R. (2003). NCBI Reference Sequence Project: update and current status. *Nucl. Acids Res.*, **31**(1), 34–7.

7. Schuler (1997). Pieces of the puzzle: expressed sequence tags and the catalog of human genes. *J. Mol. Med.*, **75**(10), 694–8.

8. McKusick, V. A. and Amberger, J. S. (1993). The morbid anatomy of the human genome: chromosomal location of mutations causing disease. *J. Med. Genet.*, **30**(1), 1–26.

9. Press, R. D. (1999). GeneClinics medical genetics knowledge base. *Mol. Diagn.*, **4**(3), 256.

10. Khoury, M. J. (1999). Human Genome Epidemiology: translating advances in human genetics into population-based data for medicine and public health. *Gene. Med.*, **1**(3), 71–3.

11. Rebhan, M., Chalifa-Caspi, V., Prilusky, J., and Lancet, D. (1998). GeneCards: a novel functional genomics compendium with automated data mining and query reformulation support. *Bioinformatics*, **14**(8), 656–64.

12. Francis S. Collins, Mark S. Guyer, and Aravinda Chakravarti (1997). Variations on a theme: cataloging human DNA sequence variation. *Science*, **278**(5343), 1580–1.

13. Sherry, S. T., Ward, M. H., Kholodov, M., Baker, J., Phan, L., Smigielski, E. M. *et al.* (2001). dbSNP: the NCBI database of genetic variation. *Nucl. Acids Res.*, **29**(1), 308–11.

14. Brookes, A. J., Lehvaslaiho, H., Siegfried, M., Boehm, J. G., Yuan, Y. P., Sarkar, C. M. *et al.* (2000). HGBASE: a database of SNPs and other variations in and around human genes. *Nucl. Acids Res.*, **28**(1), 356–60.

15. Kogelnik, A. M., Lott, M. T., Brown, M. D., Navathe, S. B., and Wallace, D. C. (1998). MITOMAP: a human mitochondrial genome database—1998 update. *Nucl. Acids Res.*, **26**(1), 112–15.

16. Krawczak, M., Ball, E. V., Fenton, I., Stenson, P. D., Abeysinghe, S., Thomas, N. *et al.* (2000). Human Gene Mutation Database—a biomedical information and research resource. *Hum. Mutat.*, **15**(1), 45–51.

17. Robinson, J., Waller, M. J., Parham, P., de Groot, N., Bontrop, R., Kennedy, L. J., Stoehr, P., and Marsh, S. G. (2003). IMGT/HLA and IMGT/MHC: sequence databases for the study of the major histocompatibility complex. *Nucl. Acids Res.*, **31**(1), 311–14.

18. Benjamin, I., Goldwein, J. W., Rubin, S. C., and McKenna, W. G. (1996). OncoLink: a cancer information resource for gynecologic oncologists and the public on the Internet. *Gynecol. Oncol.*, **60**(1), 8–15.

19. Schaefer, C., Grouse, L., Buetow, K., and Strausberg, R. L. (2001). A new cancer genome anatomy project web resource for the community. *Cancer J.*, **7**(1), 52–60.

Chapter 4

Agricultural biotechnology

Victoria Carollo

United States Department of Agriculture, Agricultural Research
Service, 800 Buchanan Street, Albany CA 94710, USA.

1 Introduction

It is widely agreed that humans made the transition from a nomadic hunter-gatherer lifestyle to one dependant largely on agriculture about 10,000 years ago in the Pre-Pottery Neolithic Near East. Archeological evidence of the Neolithic founder crops (einkorn wheat, emmer wheat, barley, lentil, pea, bitter vetch, chickpea, and flax) has been traced to a small area within the 'Fertile Crescent'—near the upper Tigris and Euphrates rivers in what today is southeast Turkey and northern Syria (1).

In the past thousand years, other plant foods introduced from Asia (rice, citrus, stone fruit, spices) and the 'New World' (maize, Solanaceous crops such as tomato, pepper, eggplant, squash) have advanced the human diet to what most of us experience today as a dazzling array of food choices in our grocery stores. Regional cuisines have blossomed, largely with introduced crops that thrive in a similar climate to where they were first domesticated.

Throughout this era of 'modern agriculture' (see *Table 1*), farmers and professional plant breeders have been selecting plants with desirable qualities making constant improvement in the yield potential, disease resistance, and nutritional quality of crop plants. In the 20 years since the first crop was genetically engineered, biotechnology and the Internet have both become household words.

Today, more than 2 trillion metric tons of crops are produced worldwide (see *Table 2*), and with the human population projected to be 9 billion people by 2050 (10), the world crop production will have to increase accordingly. Biotechnology offers plant breeders and plant biologists a powerful mechanism for crop improvement.

As plant genomes are sequenced and analysed, we will gain insight into the underlying genetic mechanisms plants use to survive drought, resist disease and

Table 1 Timeline of Agricultural Biotechnology and Internet Milestones[a]

1862	USDA created
1866	Gregor Mendel lays the foundation of modern genetics
1912	Thomas Hunt Morgan announces his theory of genes
1938	*Bacillus thuringiensis* first sold as insecticide
1953	Watson and Crick describe double-helical structure of DNA
1961	First semidwarf cultivar of a cereal grain released launching the 'Green Revolution'
1965	Lawrence Roberts and Thomas Merrill build first wide-area computer network (3)
1972	First recombinant DNA molecule synthesized
1980	Term 'ransgenic' coined
1982	First genetically engineered crop plant (tomato) developed
1983	PCR invented
1983	Transposons rediscovered
1984	First transgenic sheep and pigs born
1988	First patent for a genetically engineered animal
1990	'Gene gun' first used to genetically transform plants
1991	World-Wide Web (WWW) released by CERN (4)
1993	United Nations goes online (4)
1994	Calgene releases 'Flavr Savr' after the FDA approval (5)
1996	Yeast genome sequenced (6)
1997	Dolly cloned from the udder cell of an adult sheep
2000	*Arabidopsis thaliana* genome sequenced (7)
2000	StarLink found in taco shells
2001	Rice genome sequenced (8)

[a]All timepoints are from (2) unless where noted.

Table 2 World Crop Production Since 1998 (Million Metric Tons)(9)

Year	Wheat	Coarse grains[a]	Rice (milled)	Oilseeds[b]
1998/99	588.2	890.0	394.2	294.1
1999/00	586.8	877.4	408.4	302.8
2000/01	582.1	859.7	397.9	313.4
2001/02 (preliminary)	579.8	891.2	397.1	323.7
2002/03 (projected)	566.9	861.6	382.1	326.4

[a] Coarse grains include corn, sorghum, barley, oats, rye, millet and mixed grain.
[b] Oilseeds include soybean, cottonseed, peanut, sunflowerseed, rapeseed, copra and palm kernel.

pests, and tolerate salinity in the environment, just to name a few stresses that crops endure. Discovery of genetic factors contributing to agronomically favourable phenotypes of elite cultivars will also benefit plant breeders focusing on crop improvement through traditional breeding methods.

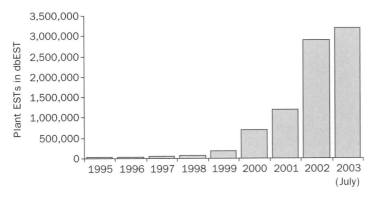

Figure 1 Plant expressed sequence tags (ESTs) that have been submitted to the dbEST database at NCBI.

2 Internet resources for agricultural biotechnology

2.1 National Center for Biotechnology Information

The National Center for Biotechnology Information (NCBI), established in 1988, is by far the most useful and complete source of molecular biology resources on the Internet. Although the focus of NCBI has been the human genome and 'understanding molecular processes affecting human health and disease', NCBI holds a vast amount of information pertinent to plant scientists as well.

Nucleotide databases at NCBI include the non-redundant (nr) database, containing gene sequences with known biological function as well as longer sequences (BACs, YACs, etc) from genomic DNA sequencing efforts; the expressed sequence tag (EST) database (dbEST) containing 'shot-gun' sequencing results generated by sequencing one end of a random cDNA clone; and the more recently added UniGene database to 'automatically partition GenBank sequences into non-redundant sets of gene-oriented clusters.' Each UniGene represents a unique gene with records linking to other tools in GenBank such as the MapViewer and Expression data. The UniGene option in the Entrez search engine provides a straightforward method to find individual gene sequences from species represented in this experimental system.

Since 1998, Plant EST submissions in the dbEST have increased dramatically (see *Figure 1*) (D. Lipshultz, personal communication), as well as the number of plant species with available sequence information (see *Figure 2*) (J. Weisemann, personal communication). At the time of this writing, of the over more than 175 plant species with sequence information in dbEST, 75 had over 1000 EST sequences available (see *Table 3*).

Other plant resources at NCBI are included within the Taxonomy Browser. From the 'Taxonomy' link at NCBI home page, one gets to the NCBI Taxonomy Homepage. This page contains direct links to common organisms, and a link

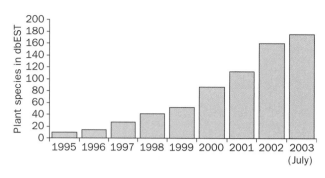

Figure 2 Number of plant species with (ESTs) submitted to dbEST by year.

Table 3 Major plant EST accessions in dbEST (>1000)

Plant	No. ESTs
Cereals	
Triticum spp. (wheat)	437,768
Hordeum spp. (barley)	348,181
Zea mays (maize)	234,019
Oryza sativa (rice)	201,794
Sorghum bicolor (sorghum)	130,106
Secale cereale (oat)	9,194
Triticum turgidum (durum wheat)	3,003
Legumes	
Glycine max (soybean)	338,122
Medicago truncatula (alfalfa)	185,621
Fruits and Vegetables	
Lycopersicon esculentum (tomato)	150,228
Vitis vinifera (wine grape)	102,420
Solanum tuberosom (potato)	94,473
Lactuca sativa (lettuce)	68,188
Capsicum annum (pepper)	22,433
Trees	
Pinus taeda (loblolly pine)	72,665
Populus spp. (poplar)	56,013
Cryptomeria japonica (Japanese ceder)	7,127
Other	
Arabidopsis thaliana (thale cress)	188,778
Gossypium spp. (cotton)	52,894
Brassica napus (canola)	37,108
Lotus japonicus (lotus)	36,262
Mesembryanthemum crystallinum (ice plant)	25,803
Nicotiana tabacum (tobacco)	9,911

to a taxonomic tree with the most current phylogenetic classification scheme. Users will find taxonomy of higher plants by clicking on 'Eukaryota' within the group of organism classifications under 'Taxonomy browser'. From the Eukaryota page select 'Viridiplantae (green plants)'—'Embryophyta (land plants)'— 'Spermatophyta (seed plants)'. Species records contain links to the genetic and mitochondrial genetic codes, the Genome view of mapped chromosomes, the taxonomic lineage, and a query form to search that species within NCBI's Entrez, a search and retrieval system that queries the nucleotide and protein sequence databases and MEDLINE articles via PubMed.

2.2 The Institute for Genomic Research

The Institute for Genomic Research (TIGR) was founded in 1992 as a non-profit research institute. Their primary research interests are in the structural, functional, and comparative analysis of genomes and gene products of a wide variety of organisms including eukaryotes, eubacteria, archaea, and viruses.

TIGR's website (see *Table 5*) is a portal for downloading a vast collection of software such as Glimmer (gene finding software), and the TIGR Array Viewer (analysis of microarray expression data), Repeatfinder (mining and analysis of repetitive sequences), the TIGR Assembler (contig construction) etc. The TIGR staff also curates several databases, including data collections for *Arabidopsis thaliana*, rice and potato. These typically include sequencing and *in silico* genetic mapping.

One of the most useful features at TIGR is the 'TIGR Gene Indices'. These indices integrate data for individual species from international sequencing efforts (genomic and EST) as well as gene research projects. Currently there are indices available for several animals, plants, protests, and fungi. Within a Gene Index users can perform searches using basic local alignment search tool (BLAST) (12), perform 'electronic northerns' by searching Gene Expression Data, or search by the gene product name. Successful searches result in a tentative consensus (TC) sequence with displays of the sequence, a contig map of the ESTs contributing to the consensi, a list of ESTs with links to individual EST sequences and GenBank records, and a list of 'Best hits' within GenBank to assign a putative identification. TCs that have been mapped will have a link within the results page to the 'alignment map' display.

Another very useful feature at TIGR is the Eukaryotic Gene Orthologs (EGO). This database 'is generated by pairwise comparison between the TC sequences that comprise the TIGR Gene Indices from individual organisms.' EGOs can be accessed by searching by TCs, gene names, accession numbers, and via a BLAST interface. For example, if a user enters 'enolase' in the 'search by gene name' query box, the result will include several 'tentative orthologs' arranged from the largest to smallest group. By clicking on the link with the ortholog number, the underlying software generates a list of ortholog members (in this case, enolase from several different organisms), and a graphical display that highlights relatedness of these sequences among the species, a sequence alignment of all ortholog members, and a link to view these results with Jalview, a multiple alignment editor.

2.3 National Genetic Resources Program/Germplasm Resource Information Network website

The US Congress established the National Genetic Resources Program (NGRP) in 1990 to 'acquire, characterize, preserve, document, and distribute to scientists, germplasm of all lifeforms important for food and agricultural production.' The US Department of Agriculture's Agricultural Research Service (USDA-ARS) maintains the Germplasm Resources Information Network (GRIN) web server to disseminate information about the NGRP holdings, which include over 10,000 species in over 1,500 genera and 454,000 accessions.

Protocol 1.1

Browsing the National Plant Germplasm System (NPGS) website

1 Point your browser to the NGRP website (www.ars-grin.gov).

2 Click on 'Plants' to go to the NPGS site.

3 Click on 'Collections' for a page that includes a map of germplasm repositories in the United States.

4 Click on 'List of Germplasm Repositories' for a page of repository addresses, species list and websites.

5 Go 'Back' and click on 'Diversity of Species' for a current summary of the NPGS holdings.

6 Go 'Back' and click on 'Summary Statistics'. From here you can browse a selection of main crops, alphabetical listings of genera and repository sites.

7 Click on 'Grape' in the Fruits column of the Common Crops table. This takes the browser to a page listing available *Vitis* species, number of accessions and countries holding germplasm.

8 Scroll down and click on *Vitis vinifera* to go to the 'Taxon' page for the species. This page lists nomenclature data, economic importance of the crop, distribution range, and references.

9 To browse through available accessions at the U.C. Davis National Germplasm Repository, click on the 'DAV' link to go to the database page for the repository site.

10 Click on 'Species held at site' to bring up a list of all holdings.

11 Scroll down and click on the link following *Vitis vinifera* to bring up a list of cultivar accessions at U.C. Davis.

12 Click on any of the cultivar names for the accession information page. Information will include the form the germplasm was received, source history, and availablity.

Protocol 1.2

Querying the NPGS website

1 Point your browser to the NGRP website (www.ars-grin.gov).

2 Click on 'Plants' to go to the NPGS website.

3 Click on 'Search GRIN' for a list of database query options.

4 Click on 'Accession Area Queries' for a query page allowing users to ask from simple to complex questions of the database.

5 Enter 'Bacchus' in the 'Simple queries' text box, and click on the 'Submit Simple Query" button to bring up list of available accessions at all repository sites.

6 Go 'Back' to the 'Search the GRIN System' page and select 'Taxonomic Queries".

7 Select 'Simple queries of species' data, and enter 'Bacchus'. Since this is a cultivar, and not a species there will be no results.

8 Go 'Back' and enter a valid species name, country in the species range, or economic use. For example, if one were to search the term 'weed', a list of weedy species would result.

9 Go 'Back' to the GRIN Taxonomy page and click on 'Economic Plants' under Specialized Queries.

10 Here, users can construct queries by selecting economic impact classes and subclasses.

11 Select the 'Beverage Base' subclass in the 'Food' class and 'Submit Query' to bring up a list of plants used in beverage making.

2.4 PlantGDB—Plant Genome Database

The Plant Genome Database was funded in 1999 by the NSF to '[1] to set up annotated species-specific EST databases for all major plants and link the EST data to other biological information, [2] to establish inter-species query capabilities to allow cross-species comparisons, and [3] to provide Web-accessible computational tools for gene identification and characterization'.

Individual species catalogued at PlantGDB are accessed via a taxonomic tree or list of accessions that include grains, cotton, tomato, maize, rice, alfalfa and pine. Species Web pages provide a general description of the plant, links to crop-specific databases and genome projects, and links to tools maintained at the PlantGDB site including a BLAST search engine for sequence queries, a text search query tool and a link to GeneSeqer (13). GeneSeqer is a Web service at PlantGDB to identify potential exon/intron structure within pre-mRNA via splice-site prediction and sequence alignment of genomic and EST sequences.

The PlantGDB project is also involved in generating clusters of plant ESTs. Clusters are derived from assembling ESTs and sorting them into contigs containing two or more ESTs and singlets, which are not significantly similar to any other EST in the assembly. The collection of contigs and singlits represent unique genes, and are called TUGs (Tentative Unique Genes) by PlantGDB. These are essentially the same as UniGenes at NCBI. These TUGs are available as query targets in the GeneSeqer tool.

Protocol 2.1 briefly follows an example also presented in the GeneSeqer tutorial at PlantGDB to identify putative genes in a sorghum BAC clone representing approximately 142,000 nucleotide bases of genomic sequence. Users are urged to visit this site to learn more about refining their analysis and aligning protein sequences.

Protocol 2.1

Navigating the GeneSeqer at PlantGDB

1　Point your browser to the PlantGDB Web site (www.plantgdb.org).

2　Click on GeneSeqer@PlantGDB from the top menu.

3　STEP 1: Select splice site model. Select 'maize' from the drop-down menu, since we will be using a sorghum (a close relative to maize) BAC clone to find putative sequences.

4　STEP 2: Input Genomic DNA sequence. Users can cut and paste or upload sequences in GenBank format from their own computers. A GenBank accession number can simply be entered as well. For this example, type 'AF503433' into the query box for the GenBank accession number.

5　STEP 3: Select or input cDNA/EST sequences. Users can input their own sequences to query against the target genomic clone entered in STEP 2, or they can select from plant ESTs archived at PlantGDB. For finding regions containing known plant genes, select the TUG (tentitive unique gene) option in the 'All Plants' category.

6　STEP 4: Submit job. Click on the 'HTML-formatted output' option and the 'Submit' button. For large files, this may take several minutes. Users may opt to have the results e-mailed.

7　Users will see a page claiming to 'display the current partial job output'. Reload frequently to continue to update the page. The download is complete when a graphic appears at the top of the file.

8　Observe the Prediction Summary graphic. These are divided into 60,000 bp segments but users can select other segments at the drop-down menu on the left menu bar. A legend describing the color scheme is available when users mouse over the words 'Prediction Summary' title above the graphic, however this only appears to work in Internet Explorer. In short, arrows are predicted gene structures, rectangles are exons and lines between rectangles are introns.

9　Scroll through the alignment file at the lower pane of the page. This displays the alignment between the genomic and EST sequences and reports quality scores.

10 Click at the link on top of the lower pane stating 'Click here to access the result summary and navigation page'. This summary reports the species and GenBank accessions of matching sequences, alignment statistics, and information about the library from which the matching EST was cloned.

2.5 GrainGenes

GrainGenes, the Triticeae genome database, has been the most comprehensive source of molecular and phenotypic information on wheat, barley, rye, and oats, since its inception in 1992 (see *Figure 3*). GrainGenes is funded by USDA-ARS and is curated as a joint effort at the USDA-ARS Western Regional Research Center (WRRC) in Albany, CA and the ARS Genome Database Resource, at the USDA- ARS Center for Agricultural Bioinformatics, Cornell University (Ithaca, NY).

The GrainGenes website is curated at the WRRC and is the starting portal for the database 'proper'. The website incorporates a large amount of additional

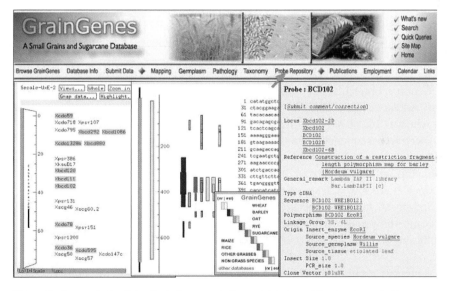

Figure 3 Screenshots of the GrainGenes website and database. The top frame is the menu bar from the web page. Clicking on 'Browse GrainGenes' or the GrainGenes logo accesses the database 'proper'. The lower left frame is an example of an interactive genetic map in the database. Clicking on any of the loci will bring up a locus text record from the database. The centre frame is a database sequence record with homology bars. These bars represent similarity of the current sequence to sequences reported for other plant species. The width of the bar implies degree of homology. The inset of the centre frame is the colour code to interpret the species with the homologous sequences. The right frame is a typical probe record from the database, with links to sequence records, references, the source of the probe, etc.

65

information relevant to the grains research community, is active in the electronic publication of several newsletters, and maintains up to the minute grains-related news, employment opportunities, and meeting information.

The GrainGenes AceDB database includes information on:

(a) Genetic maps: interactive map displays are generated dynamically from the AceDB database software. GrainGenes has a collection of over 500 genetic maps that represent chromosomes from many of the small grain cereals. These maps have been generated from several research laboratories on numerous germplasm accessions.

(b) DNA Probes: cloning and mapping information including loci, nucleotide sequences, PCR primers, amplification conditions, and contact sources.

(c) Genes: online publication of the Wheat Gene Catalogue, and incorporation of known Triticeae genes into the database including information on synonyms, alleles, and mapping.

(d) Germplasm: genotypes and pedigrees of cultivars, genetic stocks, and wild accessions.

(e) Pathology: descriptions of plant diseases of cereals including many full-colour images.

(f) Colleagues: addresses and research interests of cereals researchers.

(g) References: over 10,000 relevant bibliographical citations.

GrainGenes also serves as a portal for information generated by the U.S. Wheat Genome Project, a multi-agency collaboration to sequence all of the genes in *Triticum aestivum* (bread wheat), a hexaploid species. Since the genome size for wheat is 16,000 Mb (compared with 140 and 400 Mb for *Arabidopsis* and rice, respectively) sequencing the entire genome at this time would be prohibitively expensive and time consuming. The Wheat Genome Project therefore is sequencing only ESTs derived by sequencing one end of a random cDNA clone. Since there are at least 25,000 genes expressed during plant development, finding all genes, especially ones that are rarely expressed is a large undertaking; the project generated ESTs from over 40 distinct libraries made from different tissues, at different developmental stages, and under various stresses.

EST sequences are immediately compared with known sequences in GenBank via a BLAST search, thus providing a rapid way to identify the potential function of the gene. This information is then posted to the project website providing a nearly instantaneous flow of information from the laboratory to the public via the Internet.

GrainGenes also assists the administration of ancillary Triticeae genomics projects including simple sequence repeats (SSRs, a.k.a. microsatellites) coordination efforts; single nucleotide polymorphism (SNP) development projects; TREP the Triticeae Repeat Sequence Database; marker-assisted selection; and physical mapping projects.

2.6 MaizeDB

MaizeDB is a comprehensive source of information on the molecular biology and genetics of maize. It has been supported by the USDA-ARS since 1991, and more recently also with funds contributed by the National Science Foundation (NSF) and the University of Missouri. MaizeDB is curated at the University of Missouri (Columbia) but is planning to merge with MaizeGDB in late 2003.

MaizeDB has a very well integrated web site and database that runs on Sybase relational database software. Among the features on the web site are an extensive image collection, maize maps, germplasm resources, and links to several different search strategies for querying the database.

The MaizeDB Simple Sequence Repeats (SSRs, aka microsatellites) web page contains tables of SSR markers developed from known genes (many with known mapping coordinates), PCR primer pairs to amplify regions containing the SSRs, and map displays of chromosomes with mapped SSR loci. This feature of MaizeDB is the top hit for the web site (M. Polacco, personal communication).

A recent addition to MaizeDB is the comparative mapping tool (see *Figure 4*). This viewer was developed as collaboration between the Maize Mapping Project at Columbia, MO, USA and the Rice Genome Program in Tsukuba, Japan. Users are presented with a form where they can select among chromosomes of several mapping populations. The result is a graphical map browser with loci linked

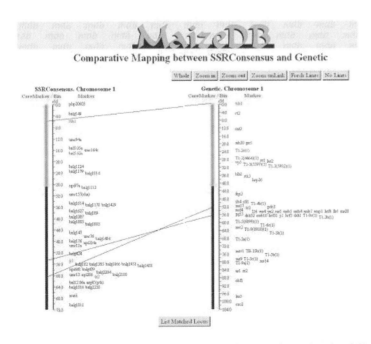

Figure 4 Screenshot of the MaizeDB comparative map viewer (version 1.0). This example is comparing Chromosome 1 from two mapping populations. Each chromosome diagram is accompanied by a centiMorgan scale, and loci are listed to the right. Zoom features allow users flexibility in viewing the maps. Lines connect matched loci.

to several databases including GenBank, SwissProt, the DNA DataBank of Japan (DDBJ), GrainGenes and RiceGenes.

The Interactive Maize Plant (IMP), while still under construction at the time of this writing, is an excellent resource for plant biology and plant anatomy education. Similar to the body part viewer at FlyBase, an image map of an entire maize plant is shown, and when the user positions the mouse over parts of the plant (currently only the tassel or a lateral branch with an ear), a detailed diagram appears with anatomical terms listed within a description of that plant part. Placing the mouse over individual anatomical parts will highlight the anatomical term and reciprocally, mousing over a term will highlight the anatomical feature in the illustration.

3 Plant genomics and additional agricultural biotechnology web resources

Two of the hottest topics in agricultural biotechnology today are genomics and genetically modified organisms (GMO) foods. Genomics is the study of genomes (all genes in an organism) and relies heavily on high-throughput laboratory equipment (robotics for colony picking, liquid handling, arraying, and multi-sample sequencers) and the Internet to rapidly gather, analyse, and disseminate data.

Other '-ome' terms coined recently are:

- Transcriptome—the 'expression profile', or population of mRNAs in a cell at a given time. Given recent evidence of alternative splicing sites in mRNA molecules (7), the transcriptome is rapidly moving to the forefront of research interests.

- Proteome—all proteins made in a cell.

- Metabolome—all small molecules in a cell including salts, sugars, amino acids, and nucleotides, including molecular intermediates and breakdown products.

Genomics requires extensive computational resources and there are several excellent sources of free software available on the web (see *Table* 4). Several other excellent web sites for agricultural biotechnology are listed in *Table* 5.

GMOs have been genetically engineered to express a foreign gene. In traditional breeding, wide crosses were made among plants or animals to bring a gene of interest, or desirable trait, from a distantly related organism into an elite germplasm. This was subsequently followed by several generations of backcrossing to reestablish a plant or animal with the desirable traits of the elite germplasm, and the expression of the new gene. Using biotechnology, specific genes can be introduced into organisms, considerably hastening the time required to release an improved germplasm. Biotechnology promises to greatly enhance the quality of food, however, this topic is surrounded by a great deal

Table 4 Free software sources for genome research

Name	Type	URL
Phred/Phrap/Consed	Base-calling and assembly	`http://www.phrap.org`
Crossmatch	Vector and repeat masking	`http://www.phrap.org`
TIGRs suite of tools	Gene finding, Alignment, Sequence Finishing, Microarray Tools, and more	`http://www.tigr.org/software`
European Bioinformatics Institute suite of tools	Vast selection of online and downloadable tools	`http://www.ebi.ac.uk/services`
Staden Package	Sequence assembly	`http://www.mrc.lmb.cam.ac.uk/pubseq/`

Table 5 Other web resources for agricultural biotechnology

Genome databases	
National Genetic Resources Program (NGRP/Grin)	`http://www.ars-grin.gov`
National Center for Biotechnology Information (NCBI)	`http://www.ncbi.nih.gov`
PlantGDB — Plant Genome Database	`http://www.plantgdb.org`
The Institute for Genome Research (TIGR)	`http://www.tigr.org`
UK CropNet	`http://ukcrop.net/`
Plant genome project Websites	
The Arabidopsis Information Resource (TAIR)	`http://arabidopsis.org`
Clemson University Genomics Institute (CUGI)	`http://www.genome.clemson.edu`
GrainGenes	`http://wheat.pw.usda.gov`
MaizeGDB	`http://www.maizegdb.org`
MaizeDB	`http://www.agron.missouri.edu`
North American Barley Genome Project	`http://barleyworld.org/NABGMP/nabgmp.html`
Pig and Chicken Genome Mapping Project	`http://www.genome.iastate.edu`
Rice Genome Research Program	`http://rgp.dna.affrc.go.jp`
U.S. Meat Animal Research Center	`http://www.marc.usda.gov/`
U.S. Wheat Genome Project	`http://wheat.pw.usda.gov/NSF`
Websites for major centers in plant biology	
Cold Spring Harbor Laboratory	`http://www.cshl.org`
CIMMYT — International Maize and Wheat Improvement Center	`http://www.cimmyt.org/`
International Rice Research Institute (IRRI)	`http://www.irri.org`

Table 5 *(Continued)*

Institute of Plant Genetics and Crop Plant Research	http://pgrc.ipk-gatersleben.de
John Innes Centre	http://www.jic.bbsrc.ac.uk
Scottish Crop Research Institute (SCRI)	http://www.scri.sari.ac.uk
Agricultural biotechnology and GMO-food resources	
APHIS—Animal and Plant Health Inspection Service	http://www.aphis.usda.gov
Biotechnology—An Information Resource	http://www.nal.usda.gov/bic
Council for Biotechnology Information	http://www.whybiotech.com
U.C. Berkeley Outreach	http://ucbiotech.org
U.C. Davis Biotechnology	http://www.biotech.ucdavis.edu/
Dupont Biotechnology	http://www.dupont.com/biotech
Monsanto	http://www.monsanto.com
Monsanto—food biotechnology in the UK	http://www.monsanto.co.uk
Union of Concerned Scientists	http://www.ucsusa.org
Plant biology education websites	
American Society of Plant Biologists (ASPB) Education Foundation	http://www.aspb.org/education/
National Agricultural Library Kids' Science Page	http://www.nal.usda.gov/Kids
The Oregon Wolfe Barleys	http://barleyworld.org/Wolfebar/Wolfnew.htm
Other plant-related web resources	
Curtis Botanical Magazine	http://www.nal.usda.gov/curtis
National Agricultural Library	http://www.nal.usda.gov
United States Department of Agriculture (USDA)	http://www.usda.gov
USDA Plants Database	http://www.usda.gov/plants
Vascular Plant Image Gallery	http://www.csdl.tamu.edu/FLORA/gallery.htm
Wayne's Word of Unusual and Noteworthy Plants	http://waynesword.palomar.edu/worthypl.htm
Link Libraries	
National Agricultural Library Plant Genome Information Resource	http://www.nal.usda.gov/pgdic/
Bioinformatics Resources on the World Wide Web	http://zlab.bu.edu/zlab/links
The Plant Link Library	http://www.dpw.wageningen-ur.nl/links/

of controversy. Recent consumer backlash over the transgenic maize carrying a Bt gene (Starlink) that was not yet approved for human consumption, yet was detected in maize products intended for humans, has cast doubt on the willingness of consumers to readily accept biotech foods. Issues of labelling genetically modified foods, rights of corporations to patent genes, and the large scale planting of transgenic crops in which consumers can see little direct benefit for them, that is, longer shelf-life, more nutritious, etc., all add to the growing concern over our food source.

Web-based information about GMO foods at this time is limited to websites from academic, corporate, and independent institutes (see *Table 5*). In the United States, the National Research Council (NRC), the research arm of The National Academies, has called upon the USDA, the Environmental Protection Agency (EPA) and the Food and Drug Administration (FDA) to establish a GMO research database (14). The USDA's Animal and Plant Health Inspection Service (APHIS) hosts a web site that describes their role in the regulation of movement, importation and field testing of genetically engineered plants and animals, and is an excellent source for information on the USDA's role in agricultural policy concerning GMO issues.

References

1. Lev-Yadun, S., Gopher, A., and Abbo, S. (2000). *Science*, **288**, 1602.
2. Agricultural Research Service (2001). *ARS Timeline.* http://www.ars.usda.gov/is/timeline/index.htm (8 August 2001).
3. Leiner, B. M. *et al.* (2000). *Brief history of the Internet*, v.3.31. http://www.isoc.org/internet/history/brief.html (9 August 2001).
4. Zakon, R. H. (2001). *Hobbes' Internet Timeline*, v.5.3. http://www.zakon.org/robert/internet/timeline/ (9 August 2001).
5. Stone, B. (1994). *Biotechnology Information Center.* http://www.nalusda.gov/bic/Federal_Biotech/news/1994.news/flavr.fda.html (13 August 2001)
6. Goffeau, A. *et al.* (1996) *Science*, **274**, 546.
7. The Arabidopsis Genome Initiative (2000). *Nature*, **408**, 796.
8. Barry, G. F. (2001) *Plant Physiol.*, **125**, 1164.
9. FASonline (2001). *World agricultural production crop production tables.* http://www.fas.usda.gov/wap/circular/2001/01-03/tables.html (10 August 2001).
10. U.S. Census Bureau (2000). *World population information.* http://www.census.gov/ipc/www/worldpop.html (13 August 2001).
11. TIGR (2001). *TIGR The Institute for Genomic Research.* http://www.tigr.org/about/ (8 Aug 2001).
12. Altschul, S. F., Gish, W., Miller, W., Myers, E. W., and Lipman, D. J. (1990). *J. Mol. Biol.*, **215**, 403.
13. Usuka, J., Zhu, W., and Brendel, V. (2000). *Bioinformatics*, **16**, 203.
14. Macilwain, C. (2000). *Nature*, **404**, 693.

Chapter 5

Inference and applications of molecular phylogenies: An introductory guide

Korbinian Strimmer* and
David L. Robertson[†]
*Department of Statistics, University of Munich, Ludwigstr. 33,
80539 Munich, Germany.
[†]School of Biological Sciences, University of Manchester, Oxford
Road, Manchester MI3 9PT, UK.

1 Introduction

Molecular phylogenies have a wide range of practical applications in the analysis of DNA sequences and are now an essential tool in areas ranging from population genetics to genomics to virology. The use of modern computers has fostered the development of sophisticated methodologies, and subsequently a large number of programs have become available. In this chapter we give an introductory overview of the most important methods of inferring phylogenetic trees from nucleotide or amino acid sequence data, emphasizing concept rather than mathematical detail (*Box 1*). We discuss simple guidelines for choosing a method appropriate for a data set. Advanced issues in molecular phylogenetics are briefly introduced, along with suggestions for further reading. The versatility of molecular phylogenetics is also highlighted and we present a range of practical problems where evolutionary trees have formed a key component of the analysis. Finally, we provide an overview of phylogenetic software and corresponding Internet resources (*Box 2*).

Box 1 Steps for inferring a phylogenetic tree

This chapter deals with methods (main text) and programs (*Box 2*) to infer phylogenetic trees from DNA sequences. The following are the basic steps for preparing the data for phylogenetic analysis and for obtaining a publication-ready output:

1 Retrieve homologous DNA or protein sequences, for example, from a public database like NCBI's GenBank (http://www.ncbi.nlm.nih.gov/) or the EMBL Nucleotide Sequence Database (http://www.ebi.ac.uk/embl/).

---

2 Inference of phylogenetic trees

A natural means to illustrate the evolutionary relationships among a sample of DNA sequences is in the form of a tree (*Figure 1*). The branching order of the tree (its topology) indicates how the sequences are related to each other, that is, which sequences share a most recent common ancestor. If included, branch lengths of a tree represent genetic distance. Evolutionary trees can be depicted as *unrooted* (*Figure 1(a) and (c)*) or *rooted* (*Figure 1(b)*).

Whenever sequence data are analysed it is important to account for the statistical dependencies between the sequences indicated by the phylogenetic tree. However, true evolutionary relationships among the sequences are rarely known. Instead, the phylogenetic tree is inferred from the data. Unfortunately, reconstructing accurate evolutionary trees from genetic data is intrinsically difficult, due to three main factors:

First, to infer a tree from DNA sequences a reliable multiple alignment must be constructed. However, an alignment in itself infers a particular evolutionary

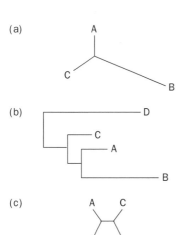

Figure 1 Unrooted (a, c) and rooted (b) phylogenetic trees for sequences A, B, C, and D. Branch lengths represent genetic distance (e.g. nucleotide substitutions).

Table 1 Number of possible tree topologies

Sequences	Unrooted trees	Rooted trees
3	1	15
4	15	105
5	105	945
6	945	10395
7	10395	135135
8	135135	2027025
9	2027025	34359425
10	34359425	654729075
15	2.13458×10^{14}	6.19028×10^{15}
20	8.20079×10^{21}	3.19831×10^{23}
50	2.75292×10^{76}	2.72539×10^{78}

history by attempting to determine which sites along the sequences are homologous. Moreover, if the alignment is unreliable the accuracy of any tree inferred from it will be compromised.

Second, the number of possible trees relating a set of DNA sequences increases very rapidly as the number of sequences increases. *Table 1* shows the number of possible rooted and unrooted tree topologies for 3–50 sequences. Even for moderate numbers of sequences the number of possible trees is extremely large, rendering it impossible to scrutinize all possible arrangements in reasonable computer time.

Third, evolutionary relationships can be very complex, and similarity between sequences is generally not a good indicator of phylogenetic relatedness. For example, in *Figure 1(a)* the shortest genetic distance is between sequences A and C, while A and B are apparently more distantly related. However, when the tree is rooted with a known outgroup (sequence D in *Figure 1(b)*) it becomes clear that sequences A and B (and not A and C) share a more recent common ancestor. The same can be observed in the unrooted tree in *Figure 1(c)* where again the genetic distance between A and C is smallest but A is nevertheless grouped with B. Therefore, except under special circumstances, simple cluster methods that rely purely on genetic similarity are not suitable for recovering evolutionary relationships.

For the above reasons, no general multi-purpose algorithm to infer evolutionary trees is known that is suitable for all kinds of data. Instead, a whole suite of complementary phylogenetic methods are commonly used, each with their particular strengths (and weaknesses). Generally the more rigorous a particular method the more computational resources it requires.

2.1 Important methods

Exhaustive studies have been performed on simulated data to investigate the accuracy and other properties of tree reconstruction methods (1–4). In the following we briefly introduce the three major classes of phylogenetic inference methods that are frequently used.

2.1.1 Maximum-parsimony

Parsimony methods were among the first methods for reconstructing phylogenetic trees, and are still widely used. The philosophy behind this approach is 'Occam's razor' applied to the distribution of substitutions along the branches of a tree. *Maximum-parsimony* aims at finding the tree that explains the observed pattern of nucleotides (or amino acids) with the smallest number of mutations possible (5). This is done by placing substitutions on possible tree topologies so as to minimize the total number of substitutions required to explain the nucleotide (or amino acid) at each branch tip. The most parsimonious reconstruction, that is, the one requiring the fewest substitutions, is chosen for each site. The total number of evolutionary changes on a tree is the sum of the number of changes for each site, and the tree topology that has the fewest changes is chosen as the most parsimonious tree. The various types of parsimony methods (6) differ only in the particular details of how the above steps are pursued.

The maximum-parsimony approach generally works quite well. However, it also has a number of pitfalls that may at times render it impractical. First, the number of most-parsimonious trees can be quite large so that there can exist large numbers of equally valid solutions. Second, as an exhaustive tree search, that is, the evaluation of all possible trees, is not feasible except for small data sets, generally no guarantee can be given that the optimal tree has been found. Third, the most-parsimonious tree is not necessarily a good explanation for the data, particularly when sequences are short (7). Fourth, maximum-parsimony does not make efficient use of all the available data because constant and other non-parsimony-informative sites are ignored (6). Fifth, parsimony is particularly prone to the 'long branches attract' problem, that is, it tends to group long branches together even if this does not reflect the true evolutionary history (8).

2.1.2 Distance methods

Distance-based methods work on pairwise genetic distances which have been computed from a sequence alignment, using a suitable *nucleotide or amino acid substitution model*, rather than analysing the sites in the alignment directly.

Neighbour-joining (9,10) is one of the most popular distance methods, primarily due to its speed at finding a single 'best' tree with large data sets. This clustering approach starts from a star-like tree and resolves the tree by iteratively joining groups of sequences together, and in each step minimizing the total sum of branch lengths. Neighbour-joining has been shown to be efficient at recovering the correct tree topology (3,4).

In contrast, *UPGMA* (11) simply clusters groups of sequences that exhibit the smallest distance between each other. As phylogenetic relationships are not necessarily linked to sequence similarity, this approach often fails to reconstruct the true tree except when the sequences have evolved in a clock-like manner, that is, it requires that the rate of evolution has been constant (3).

The *least-squares* method (12, 13) optimizes branch lengths on a tree by minimizing the error of the tree-induced pairwise distances when compared to the distances computed from the alignment. An optimal tree is then found by evaluating all possible (or all feasible) candidate trees. The least-squares approach has a solid statistical foundation and so is often considered the best distance method (14).

Minimum-evolution (15), in a way similar to the least-squares method, also uses a tree search to find the best tree. The selection criterion in this approach is the total sum of branch lengths. Thus, it can be considered as somewhat similar to maximum-parsimony but applied to distance data. Note that a neighbour-joining tree is an approximation of the minimum-evolution tree.

2.1.3 Maximum-likelihood

Maximum-likelihood (ML) (16) is the approach that is generally considered to make the most efficient use of the data and to provide the most accurate estimates of a phylogenetic tree. It is, however, also the most computationally demanding technique for reconstructing phylogenetic trees.

The basic idea of the likelihood approach is to compute the probability of the observed data assuming it has evolved under a particular evolutionary tree and a given probabilistic model of nucleotide (or amino acid) substitution. The ML tree is then the tree with the highest probability of explaining the data. Technically, the likelihood of a tree is computed using a so-called directed graphical model for the sequence data (16, 17).

Inferring evolutionary trees using the ML principle is difficult for three main reasons. First, the likelihood itself is complicated to compute, essentially due to the problem of inferring the unknown ancestral character states in a tree. Second, even for a single tree topology, estimation of branch lengths is a very difficult optimization problem. Finally, computing likelihoods for all possible tree topologies is an impossible task even for medium-sized sets of sequences so that instead a heuristic tree search must be used. Thus it is no surprise that ML tree inference has only become popular since the widespread availability of faster computers.

As a statistical approach the ML method enjoys a number of advantages, such as estimators with low bias and variance compared to other methods. Moreover, the likelihood framework provides a rigorous basis for statistical testing of competing hypothesis (18). In addition, it can be shown that ML encompasses the maximum-parsimony method, the latter being an approximation to ML (19). The ML approach can also be used to compute genetic distances between sequences, for example, as prerequisite for a distance-based approach to tree reconstruction.

2.2 Common features

To obtain a better understanding of the relationships between the discussed methods it is helpful to look at alternative classifications of phylogeny inference procedures. In *Table 2* we present a number of criteria that highlight some shared features of the various methods.

Table 2 Features of tree reconstruction methods

Method	Data		Tree search		Substitution model	
	Character	Distance	Exhaustive[a]	Clustering	Explicit	Implicit
Maximum-parsimony	$\checkmark$		$\checkmark$			$\checkmark$
Neighbour-joining		$\checkmark$		$\checkmark$	$\checkmark$[b]	
UPGMA		$\checkmark$		$\checkmark$	$\checkmark$[b]	
Least-squares		$\checkmark$	$\checkmark$		$\checkmark$[b]	
Minimum-evolution		$\checkmark$	$\checkmark$		$\checkmark$[b]	
ML	$\checkmark$		$\checkmark$		$\checkmark$	

[a]In practice a suitable heuristic shortcut will often have to be used.
[b]If model is used to compute pairwise distances.

An important characteristic is how the underlying data, that is, the sequence alignment, is used. There are two main classes: methods based on character state data where sites in an alignment are individually analysed, and methods based on distance data where a pairwise distance matrix relating all possible pairwise sequence comparisons to each other is used to summarize the information in the alignment. Methods that work directly on sequence data (e.g. ML) can be expected to be more accurate than those using distance data (e.g. least-squares). However, pairwise distances often provide a remarkably compact condensation of the major features of the data.

A second criterion which can be used to distinguish methods is the way the optimal tree is obtained. Some methods simply cluster sequences together according to a set of given rules (e.g. neighbour-joining) whereas tree-search methods use specific optimality criteria to choose among all possible tree topologies. The latter methods that involve tree evaluation (specifically maximum-parsimony, ML, least-squares, and minimum-evolution) can be implemented with a number of search strategies such as exhaustive search or branch-and-bound, which guarantee an optimal solution but also have, potentially, very large run-times. Alternatively, heuristic searches based on rearrangement of subtrees (6) or reassembling of quartet trees (20, 21) can be employed but these do not guarantee the optimal tree will be found. Stochastic tree searches based, for example, on genetic algorithms (22) or Markov chain Monte Carlo techniques (23) have also been implemented.

Another distinctive feature of phylogenetic tree reconstruction methods is the use of the model of substitution. The purpose of a substitution model is to estimate the actual amount of evolutionary change. This is problematic because there is not a simple relationship between the observed and actual changes, and the same site in a sequence might have undergone repeated substitutions while only the last substitution can be observed. Thus, any substitution model must 'correct' for these unobserved changes. In nucleotide substitution models factors such as transitions being more frequent than transversions and unequal base composition are also incorporated into the model. For detailed discussion of the available models see (6). Note that in maximum-parsimony the model of substitution is implicit in the method. This does not imply that maximum-parsimony is model-free. Rather, methods like maximum-parsimony may often be inadequate for sequence data because of its simple implicit substitution model, whereas in ML and distance methods a complex, and hopefully realistic, model can be employed. Maximum-parsimony procedures have been implemented that attempt to give different weights to different types of substitutions, for example, to transitions and transversions (6).

2.3 Selection of suitable method

When confronted with the large variety of methods available for analysing a sequence data set it can often be difficult to choose a suitable method. However, the following simple guidelines may be useful:

1 If there are only a few sequences (<10) in the data set then ML, using an exhaustive tree search, should be the method of choice. For a medium-sized data set (10–100 sequences) ML is still applicable, but heuristic tree searches have to be employed. Maximum-parsimony or least-squares are also good alternatives for this size of data set. For very large numbers of sequences (>100) distance-based methods like neighbour-joining are the only computationally feasible approaches that can be used. It is important to keep in mind that the number of sites in the alignment will also impact the runtime of a program.

2 If at all possible more than one tree reconstruction method should be employed to check whether the same branching pattern is consistently recovered.

3 Note that if sequences are very similar or highly divergent the phylogenetic signal will be weak so it is generally unlikely that the corresponding tree can be fully resolved.

4 When selecting a model of substitution in ML, or with distance methods, it is a good idea to start with a simple substitution model, and to repeat the analysis with a more complex model and to carefully observe changes in tree topology and branch lengths.

2.4 Other methods

The search for new phylogenetic methodologies and improved models of sequence evolution is a very active area of research and a wide range of phylogenetic methods have been proposed, many of which are variants of the methods mentioned before. For example, several modifications of neighbour-joining have been suggested (24–26). Research into methodologies has also focused on the more theoretical aspects of phylogenetic reconstruction, such as phylogenetic invariants (27–30), phylogenetic spectra (31,32), and phylogenetic geometry (33), and on the employment of advanced techniques like neural networks (34) or genetic algorithms (22). Much effort has also gone into developing sophisticated statistical models of substitution processes (35–37). Another research direction is the study of algorithms suitable for very large data sets (38, 39).

2.5 Phylogenetic uncertainty

So far we have given a brief overview of how a phylogenetic tree can be estimated from molecular data. In the investigation of real data this step rarely is the end of the analysis, rather often it is only the beginning.

One of the most important questions is the problem of assessing the accuracy of the obtained tree. The most commonly used method is the bootstrap approach which has been developed to evaluate the stability of internal branches in trees (40, 41). Bootstrapping repeatedly samples sites with replacement from the original alignment until a new alignment of the same length as the original is obtained. The same phylogeny method used to construct the tree from the original alignment is then applied on the sampled alignment. This procedure is repeated a set number of times (commonly 1000) and if a particular grouping of sequences is found in 75% or more of replicates those clusters are considered well-supported (42). Other methods exist that assess the phylogenetic signal of a data set prior to the inference of a tree (43).

A single 'best' tree is only a point estimate of the true evolutionary relationships. Thus, the aim to better account for phylogenetic uncertainty has led to a general trend towards multiple-tree analysis. For example, likelihood ratio tests are frequently used to compare competing phylogenetic hypotheses in the form of alternative tree topologies (44). For many data sets it is, however, more suitable to introduce the notion of confidence sets of trees, where an ensemble of trees is preferred over a single tree. The likelihood method and the related Bayesian framework (23) are particularly amenable to this kind of analysis. This is an active area of research and various tests have been developed to determine confidence sets of trees (45, 46).

Closely related to this topic are evolutionary networks which are used to represent non-tree-like evolution, that is, relationships among the sequences which cannot be represented by a tree. A large variety of different types of networks exists. As for evolutionary trees, there are parsimony (47, 48), distance-based (49) and likelihood methods (17, 50) for inferring evolutionary networks. Splitgraphs (51) merely represent incompatibilities in distance data while median networks (48) and ancestral recombination graphs (52–54) can be interpreted as collections of trees. A median network contains all the most-parsimonious trees for a data set, and an ancestral recombination graph with r recombination events combines $r + 1$ site-specific clock-like evolutionary histories.

2.6 Further reading

General text book introductions to molecular evolution and phylogenetics are provided by (55–57). An introductory guide with detail on the use of some specific phylogeny software is given in (58). Precise algorithmic details of tree reconstruction methods are reviewed in (6) whereas (59) gives a general overview of probabilistic sequence analysis. Advanced theoretical topics in phylogeny are discussed in (60).

3 Applications

Phylogenetic trees play an important role in a large variety of problems, from all fields of biology and related disciplines (61). The richness of applications illustrates the central position that molecular phylogeny has assumed in modern biology. In the following we list some interesting case studies. For further details we refer to the original papers.

3.1 Sequence classification

Trees provide natural hierarchical classifications of the investigated sequences. Consequently, this has been one of the first applications of phylogenetic trees.

1 In taxonomy, species phylogenies are now commonly inferred using reconstructed gene trees. This is particularly interesting when there are no morphological data, or when molecular and morphological data contradict. For example, the relationship between the major taxonomic kingdoms has been established entirely from molecular sequences (62). In the inference of gene trees it is important to distinguish between the actual species tree and the embedded gene trees (63, 64).

2 In virology, phylogenetic trees have been used to define significant phylogenetic clusters of viruses. These can be used to test whether patients are linked epidemiologically and to study infected populations (65).

3 The prediction of gene function can be substantially refined by taking the phylogeny of the investigated gene into account (66). This approach is called phylogenomics (67).

3.2 Analysis of population history

Phylogenetic trees also contain information about the demographic history of the population from which the sequences were sampled. This is relevant for problems in anthropology, epidemiology, and virology.

1 Population size changes over time influence the shape and the branch lengths of trees. Coalescent theory (68, 69) provides a probabilistic model for this process and thus allows the extraction of information concerning population history from an inferred tree (70, 71).

2 Trees also allow the determination of the geographic origin of a population. For example, the common ancestor of human mitochondrial DNA (mitochondrial Eve) has been shown to be of African origin (72).

3 In addition, phylogenetic trees provide information about the time-frame of evolutionary events, for example, about the age of the mitochondrial Eve (72) or the time of the human-ape split (73).

3.3 Processes in molecular and genome evolution

Evolutionary processes in genes and genomes leave phylogenetic signals in the sequence data. Consequently, molecular phylogenies are important tools, for instance, for:

1 estimating model parameters and substitution rates (37);

2 detecting recombination in the evolutionary history of sequences (47,74,75);

3 inferring duplication events in genomes (76, 77);

4 inferring gene rearrangements in and between genomes (78);

5 detecting adaptive evolution, for example, change of function and selection (79–81); and

6 comparing host and parasite phylogenies, that is, for inferring co-phylogenies (55).

3.4 Comparative studies

Comparative analysis aims at establishing correlations between traits across taxa, for example, between brain and body size (82). It is important to discriminate dependencies between the investigated traits that are introduced merely by evolution from those representing true correlation. Consequently, phylogenetic comparative methods have been developed that explicitly take the underlying evolutionary tree into account (82, 83). Phylogenetic uncertainty can also be incorporated in such an analysis (84).

3.5 Other bioinformatics applications

There are many other applications of evolutionary trees in bioinformatics and sequences analysis. For example:

1 Phylogenetic trees are important for the reconstruction of ancestral sequences (5, 85, 86).

2 To compute sequence alignments some algorithms rely on the inference of evolutionary trees (59, 87).

3 Database searches for homologous sequence relationships can be made more efficient if the phylogeny of the sequences in the database is used as guide tree (88).

Box 2 Internet resources

A large number of software packages are available to infer evolutionary trees from sequence data. The variety of this software derives directly from the richness of the available methods, with many researchers providing implementations of their own and other people's methods. An exhaustive list of phylogenetic computer programs is maintained by Joseph Felsenstein at his web page http://evolution.genetics.washington.edu/phylip/software.html. Programs dealing with recombination in the ancestry of sequences are listed at D.L.R.'s web page http://bioinf.man.ac.uk/~robertson/recombination/.

Box 2 (*Continued*)

   Currently the most widely used program for inferring phylogenetic trees from nucleotide data (using parsimony, likelihood and distance methods) is PAUP* by David Swofford (89) (`http://www.lms.si.edu/PAUP`). PHYLIP (90) is an alternative collection of programs for the analysis of nucleotide and amino acid data created by Joseph Felsenstein (`http://evolution.genetics.washington.edu/phylip`). Phylo_win by Nicolas Galtier and Manolo Gouy again implements a variety of different methods (`http://pbil.univ-lyon1.fr`). ML analysis for nucleotide and amino acid data is provided by MOLPHY (91) by Jun Adachi and Masami Hasegawa (`ftp://sunmh.ism.ac.jp/pub/molphy`) and also by TREE-PUZZLE (20) by Heiko A., Schmidt, K.S., Martin Vingron and Arndt von Haeseler at (`http://www.tree-puzzle.de`). Parsimony and minimum-evolution methods are implemented in MEGA (57) by Sudhir Kumar and Masatoshi Nei (`http://www.megasoftware.net`).

   A large selection of evolutionary models including codon-based models is implemented in PAML (92) by Ziheng Yang (`http://abacus.gene.ucl.ac.uk/software/paml.html`). A variety of useful phylogenetic specialist tools created by Andrew Rambaut are available from the web page `http://evolve.zoo.ox.ac.uk`. For Bayesian tree inference there are two packages, BAMBE (23) by Bret Larget (`http://www.mathcs.duq.edu/larget/bambe.html`) and MrBayes by John Huelsenbeck (`http://morphbank.ebc.uu.se/mrbayes/`). An object-oriented Java library for molecular evolution and phylogenetics is maintained by Matthew Goode and others (`http://www.cebl.auckland.ac.nz/pal-project/`). An R package APE for statistical analysis of phylogenetics and evolution (incl. population genetics) is maintained by Emmanuel Paradis and available from `http://cran.r-project.org`.

   Web-based servers for phylogenetic inferences are also available. A large selection of programs is offered, for example, by the server of the Institut Pasteur, Paris (`http://bioweb.pasteur.fr/seqanal/phylogeny/intro-uk.html`).

References

1. Huelsenbeck, J. P. and Hillis, D. M. (1993). *Syst. Biol.*, **42**, 247–64.
2. Hillis, D. M. (1995). *Syst. Biol.*, **44**, 3–16.
3. Huelsenbeck, J. P. (1995). *Syst. Biol.*, **44**, 17–48.
4. Strimmer, K. and von Haeseler, A. (1996). *Syst. Biol.*, **45**, 516–23.
5. Fitch, W. M. (1971). *Syst. Zool.*, **20**, 406–16.
6. Swofford, D. L., Olsen, G. J., Wadell, P. J., and Hillis, D. M. (1996). In *Molecular systematics* (ed. D. M. Hillis, C. Moritz, and B. K. Mable), pp. 407–514, Sinauer Associates, Sunderland, MA.
7. Nei, M., Kumar, S., and Takahashi, K. (1998). *Proc. Natl. Acad. Sci. USA*, **95**, 12390–7.
8. Felsenstein, J. (1978). *Syst. Zool.*, **27**, 401–10.
9. Saitou, N. and Nei, M. (1987). *Mol. Biol. Evol.*, **4**, 406–25.
10. Studier, J. A. and Keppler, K. J. (1988). *Mol. Biol. Evol.*, **5**, 729–31.
11. Sokal, R. R. and Michener, C. D. (1958). *U. Kansas Sci. B.*, **38**, 1409–37.
12. Fitch, W. M. and Margoliash, E. (1967). *Science*, **155**, 279–84.
13. Cavalli-Sforza, L. and Edwards, A. W. F. (1967). *Evolution*, **21**, 550–70.
14. Felsenstein, J. (1997). *Syst. Biol.*, **46**, 101–11.
15. Edwards, A. W. F. (1996). *Syst. Biol.*, **45**, 79–91.
16. Felsenstein, J. (1981). *J. Mol. Evol.*, **17**, 368–76.
17. Strimmer, K. and Moulton, V. (2000). *Mol. Biol. Evol.*, **17**, 875–81.
18. Goldman, N. (1993). *J. Mol. Evol.*, **36**, 182–98.
19. Steel, M. and Penny, D. (2000). *Mol. Biol. Evol.*, **17**, 839–50.

20. Strimmer, K. and von Haeseler, A. (1996). *Mol. Biol. Evol.*, **13**, 964–9.

21. Strimmer, K., Goldman, N., and von Haeseler, A. (1997). *Mol. Biol. Evol.*, **14**, 210–11.

22. Lewis, P. O. (1998). *Mol. Biol. Evol.*, **15**, 277–83.

23. Larget, B. and Simon, D. L. (1999). *Mol. Biol. Evol.*, **16**, 750–9.

24. Gascuel, O. (1997). *Mol. Biol. Evol.*, **14**, 685–95.

25. Bruno, W. J., Socci, N. D., and Halpern, A. L. (2000). *Mol. Biol. Evol.*, **17**, 189–97.

26. Ota, S. and Li, W.-H. (2000). *Mol. Biol. Evol.*, **17**, 1401–9.

27. Cavender, J. A. (1978). *Math. Biosci.*, **40**, 271–80.

28. Cavender, J. A. and Felsenstein, J. (1987). *J. Classif.*, **4**, 57–71.

29. Felsenstein, J. (1991). *J. Theor. Biol.*, **152**, 357–76.

30. Steel, M. (1994). *Appl. Math. Letters.*, **7**, 19–23.

31. Hendy, M. D. and Penny, D. (1993). *J. Classif.*, **10**, 5–24.

32. Hendy, M. D., Penny, D., and Steel, M. A. (1994). *Proc. Natl. Acad. Sci. USA*, **91**, 3339–43.

33. Kim, J. (2000). *Mol. Phylogenet. Evol.*, **17**, 58–75.

34. Dopazo, J. and Carazo, J. M. (1997). *J. Mol. Evol.*, **44**, 226–33.

35. Goldman, N. and Yang, Z. (1994). *Mol. Biol. Evol.*, **11**, 725–36.

36. Yang, Z. (1996). *Trends Ecol. Evol.*, **11**, 367–71.

37. Liò, P. and Goldman, N. (1998). *Genome Res.*, **8**, 1233–44.

38. Csürös, M. (2002). *J. Comp. Biol.*, **9**, 277–97.

39. Kim, J. (1998). *Syst. Biol.*, **47**, 43–60.

40. Felsenstein, J. (1985). *Evolution*, **39**, 783–91.

41. Efron, B., Halloran, E., and Holmes, S. (1996). *Proc. Natl. Acad. Sci. USA*, **93**, 13429–34.

42. Hillis, D. M. and Bull, J. J. (1993). *Syst. Biol.*, **42**, 182–92.

43. Strimmer, K. and von Haeseler, A. (1997). *Proc. Natl. Acad. Sci. USA*, **94**, 6815–19.

44. Huelsenbeck, J. P. and Rannala, B. (1997). *Science*, **276**, 227–32.

45. Goldman, N., Anderson, J. P., and Rodrigo, A. G. (2000). *Syst. Biol.*, **49**, 652–70.

46. Strimmer, K. and Rambaut, A. (2002). *Proc. R. Soc. London B* **269**, 137–42.

47. Hein, J. (1993). *J. Mol. Evol.*, **36**, 396–406.

48. Bandelt, H. J., Forster, P., Sykes, B. C., and Richards, M. B. (1995). *Genetics*, **141**, 743–53.

49. Bandelt, H.-J. and Dress, A. W. M. (1992). *Adv. Math.*, **92**, 47–105.

50. Strimmer, K., Wiuf, C., and Moulton, V. (2001). *Mol. Biol. Evol.*, **18**, 97.

51. Huson, D. H. (1998). *Bioinformatics*, **14**, 68–73.

52. Hudson, R. R. (1983). *Theor. Popul. Biol.*, **23**, 183–201.

53. Griffiths, R. C. and Marjoram, P. (1996). *J. Comput. Biol.*, **3**, 479–502.

54. Griffiths, R. C. and Marjoram, P. (1997). In *Progress in population genetics and human evolution*, IMA Volumes in Mathematics and its Applications, vol. 87 (ed P. Donelly and S. Tavaré), pp. 257–70, Springer Verlag, Berlin.

55. Page, R. D. M. and Holmes, E. C. (1998). *Molecular evolution: a phylogenetic approach*. Blackwell Science, Oxford.

56. Li, W.-H. and Graur, D. (1999). *Fundamentals of molecular evolution*, 2nd edn. Sinauer Associates, Sunderland, MA.

57. Nei, M. and Kumar, S. (2000). *Molecular evolution and phylogenetics*. Oxford University Press, Oxford.

58. Hall, B. G. (2001). *Phylogenetic trees made Easy: a how-to manual for molecular biologists*. Sinauer Associates, Sunderland, MA.

59. Durbin, R., Eddy, S., Krogh, A., and Mitchison, G. (1998). *Biological sequence analysis: probabilistic models of proteins and nucleic acids*. Cambridge University Press, Cambridge.

60. Mirkin, B., McMorris, F. R., Roberts, F. S., and Rzhetsky, A. (ed.) (1997). *Mathematical Hierarchies and Biology: DIMACS Workshop November 13–15, 1996*, DIMACS Series in Discrete Mathematics and Theoretical Computer Science, Vol. 37 American Mathematical Society, Providence, RI.

61. Harvey, P. H., Leigh Brown, A. J., Maynard Smith, J., and Nee, S. (ed.) (1996). *New uses for new phylogenies*. Oxford University Press, Oxford.

62. Doolittle, W. F. (1999). *Science*, **284**, 2124–8.

63. Maddison, W. P. (1996). *Syst. Biol.*, **46**, 523–36.

64. Slowinski, J. B. and Page, R. D. M. (1999). *Syst. Biol.*, **48**, 814–25.

65. Crandall, K. A. (ed.) (1999). *The evolution of HIV*. John Hopkins University Press, Baltimore.

66. Pellegrini, M., Marcotte, E. M., Thompson, M. J., Eisenberg, D., and Yeates, T. (1999). *Proc. Natl. Acad. Sci. USA*, **96**, 4285–8.

67. Eisen, J. A. (1998). *Genome Res.*, **8**, 163–7.

68. Donnelly, P. and Tavaré, S. (1995). *Annu. Rev. Genet.*, **29**, 401–21.

69. Nordborg, M. (2001). In *Handbook of statistical genetics* (ed. D. Balding, M. Bishop, and C. Cannings), pp. 179–212, Wiley, Chichester.

70. Pybus, O. G., Rambaut, A., and Harvey, P. H. (2000). *Genetics*, **155**, 1429–37.

71. Strimmer, K. and Pybus, O. G. (2001). *Mol. Biol. Evol.*, **22**, 160–74.

72. Vigilant, L., Stoneking, M., Harpending, H., Hawkes, K., and Wilson, A. C. (1991). *Science*, **253**, 1503–7.

73. Hasegawa, M., Kishino, H., and Yano, K. (1985). *J. Mol. Evol.*, **22**, 160–74.

74. Robertson, D. L., Sharp, P. M., McCutchan, F. E., and Hahn, B. H. (1995). *Nature*, **374**, 124–6.

75. McGuire, G. and Wright, F. (2000). *Bioinformatics*, **16**, 130–4.

76. Eulenstein, O., Mirkin, B., and Vingron, M. (1998). *J. Comp. Biol.*, **5**, 135–48.

77. El-Mabrouk, N., Bryant, D., and Sankoff, D. (1999). Proc. 3rd International Conference on Computational Molecular Biology (RECOMB), pp. 154–163. Pub:ACM.

78. Blanchette, M., Kunisawa, T., and Sankoff, D. (1999). *J. Mol. Evol.*, **49**, 191–203.

79. Yang, Z. (1998). *Mol. Biol. Evol.*, **15**, 568–73.

80. Suzuki, Y. and Gojobori, T. (1999). *Mol. Biol. Evol.*, **16**, 1315–28.

81. Yang, Z. and Bielawski, J. P. (2000). *Trends Ecol. Evol.*, **15**, 496–503.

82. Harvey, P. H. and Pagel, M. D. (1991). *The Comparative method in evolutionary biology*. Oxford University Press, Oxford.

83. Felsenstein, J. (1985). *Am. Naturalist*, **125**, 1–12.

84. Huelsenbeck, J. P., Rannala, B., and Masly, J. P. (2000). *Science*, **288**, 2349–50.

85. Yang, Z., Kumar, S., and Nei, M. (1995). *Genetics*, **141**, 1641–50.

86. Pupko, T., Pe'er, I., Shamir, R., and Graur, D. (2000). *Mol. Biol. Evol.*, **17**, 890–6.

87. Thompson, J. D., Higgins, D. G., and Gibson, T. J. (1994). *Nucleic Acids Res.*, **22**, 4673–80.

88. Rehmsmeier, M. and Vingron, M. (1999). *Proceedings of the 1999 German Conference on Bioinformatics*, Hannover, pp. 66–72.

89. Swofford, D. L. (1998). *PAUP*. Phylogenetic analysis using parsimony (* and other methods), version 4*. Sinauer Associates, Sunderland MA.

90. Felsenstein, J. (1993). *PHYLIP: phylogenetic inference package, version 3.5c*. Department of Genetics, University of Washington, Seattle.

91. Adachi, J. and Hasegawa, M. (1996). *MOLPHY: programs for molecular phylogenetics, version 2.3*. Institute of Statistical Mathematics, Tokyo.

92. Yang, Z. (2000). *Phylogenetic analysis by maximum-likelihood (PAML), version 3.0*. University College, London.

Chapter 6

Internet tools for cell and developmental biologists

Peter D. Vize

Department of Biological Sciences, 2500 University Drive N.W., Calgary, Alberta, Canada T2N 1N4.

The type of data generated by cell and developmental biologists is often not well suited to presentation in printed media. How can one depict a migrating cell, vesicular cycling, or embryo gastrulation with a small number of still photographs or data tables? Such dynamic processes are better suited to illustration by active means, and the Internet is an effective system for transmitting information in such forms. Dynamic data forms are not the only ones well served by the Internet—three dimensional networks are also more simple to store, visualize, and maintain using database technology. An example of 3D data sets that are well suited to Internet storage and distribution are cross-talking signal transduction pathways, where simple hyperlinks allow users to jump from pathway to pathway in multiple dimensions. This chapter discusses both general principles and specific techniques of using the Internet for the exploration of cell and developmental biology.

1 Where to find data on cell and developmental biology

It is often difficult to use general-purpose search engines to track down specific biological information on the web. Often the most efficient way to find Internet resources for cell and developmental biology are the links pages of good subject dedicated databases, and the best place to find good links pages are centralized libraries of such resources such as the Virtual Library. The virtual library is a consortium of websites that serves as the starting point for tracking down information on an enormous variety of subjects, including biology. Links pages avoid the problems mentioned above and provide a source of jumping points that can be used to begin tracking down the information you seek. Once you identify a useful link it is always worth saving it as a bookmark or as a link within your own link resource, as will be discussed below. Some excellent databases that provide links relevant to cell and developmental biology are listed in *Table 1*.

Table 1 Web Resources for cell and developmental biology

Carnegie Stages of embryo development	`http://embryo.soad.umich.edu/car nStages/carnStages.HTML`	Illustration of developmental stages, with links to resources about each stage
Flybase	`http://flybase.bio.indiana.edu/`	The Drosophila central web site. Everything you want to know about flies.
Gene Ontology Consortium	`http://www.geneontology.org`	Controlled vocabulary of gene and protein roles
Mouse 3D altas	`http://genex.hgu.mrc.ac.uk/`	3D volumetric pixel models of mouse embryos that can display tissue or gene expression domains
STKE	`http://stke.sciencemag.org/`	A growing site that aims to link all transduction pathways
Society for Developmental Biology (SDB)	`http://sdb.bio.purdue.edu/dbcinema/`	Various movies and animations, fly morphogenesis, calcium waves, filopodial dynamics, and more.
Multidimensional human embryo	`http://embryo.soad.umich.edu/`	A collection of magnetic resonance microscopy images plus optical photographs
Rotating animation of stage 16 embryo	`http://embryo.soad.umich.edu/car nStages/stage16/stage16.mov`	QuickTime movie
Morph of human development	`http://embryo.soad.umich.edu/resources/morph.mov`	QuickTime movie
WWW Virtual Library— Cell Biology	`http://vlib.org/Science/Cell_Biology/index.sHTML`	Designed for scientists, students, and educators
Virtual Library— Developmental Biology	`http://sdb.bio.purdue.edu/Other/VL_DB.HTML`	Maintained by the Society for Developmental Biology
Wormbase	`http://wormbase.org`	The C. elegans community site
Xenbase	`http://xenbase.org`	Community site for Xenopus research
Xenopus Molecular Marker Resource	`http://www.xenbase.org/xmmr/welcome.HTML`	A Xenopus site that includes morphs and movies

2 General searches

When link resources fail your next choice for tracking down information is search engines. The choice of a search engine is an extremely personal one. Each of the major search engines uses a different formula for weighting the information their spiders recover from crawls through the Internet. Title words, words inside 'meta' tags, the first 20 words on a page, multiple usage of the one word—each engine uses a set of rules to evaluate and rank information gleaned from its wanderings on the Internet. The search engine that best suits you as a user will be the one that returns the information you seek with the highest possible ranking. The way you personally select search words must match the methods that the search engine uses to rank sites. The best approach is simply to experiment until you find an engine that effectively returns the information that you seek in searches. Concept-based engines such as Excite are particularly useful. Such engines look at not only the keyword but also surrounding words to determine relevance. An example would be the word 'heart'. In our context a search including the word heart would be very different from someone looking for romance. A context-based engine looks at surrounding words, for instance 'pulmonary', 'vascular', etc., and assigns a relevancy based on this information. This greatly improves the accuracy of searches. Some of the major engines are listed in *Table 2*.

Other options for searching include meta search engines. These function by searching the search engines, rather than by indexing websites themselves. A meta-search will evaluate and rank indexes at many different search sites. Such searches are extremely broad and much more likely to bring in less obvious information, though their very broadness can also be a drawback in that extraneous information is also retrieved. Examples of meta-search engines can also be found in *Table 2*. Some browsers or operating systems have packaged engines that also search multiple search engine indices and rank results. Two examples are the OmniWeb browser tool from Omni and Sherlock from Apple.

If you run your own website, adding a search engine will help visitors track down information and make your site much more useful. The easiest way to include a search function is to use one of the free commercial services. Such services will spider (automatically search and index web page, including pages linked to your page by hyperlinks) your site on request and store an index of keywords. They generally provide some HTML or javascript code for you to cut and paste into your web page to generate the search box. These services are generally free in return for your users being subjected to advertisements that will be present when search results are returned. An alternative is to install your own search engine. If you are using a UNIX or NT server the 'htdig' search engine is an effective and free engine that can easily be installed from pre-compiled binaries (see *Table 2*).

3 Building simple but powerful web tools

As a scientist, you probably need your own links page. While a good links page that somebody else has made may serve many of your purposes, it is very unlikely

Table 2 Web searching

Search engines		
AltaVista	`www.altavista.com`	Advanced search lets you construct Boolean queries, sort results by various criteria, etc.
Google	`www.google.com`	Results are ranked by the number of other sites that link to that page, a surprisingly effective method of focusing on the more useful sites.
Yahoo	`www.yahoo.com`	
Metasearch		
Sherlock	`http://www.apple.com/ macosx/jaguar/sherlock.HTML`	Included in Mac OS X
OmniWeb browser	`http://www.omnigroup.com/ applications/omniweb/`	A commercially available web browser for Mac OS X
Datamonster	`www.datamonster.com`	Web-based meta-search engine that compiles results from various search sites
Create bookmarkable search URLs		
AutoPOST	`http://www.io.com/~jsm/ autopost/`	Attempts to add a GET-method query string to a search form URL. (see *Protocol 1*)
frmget bookmarklet	`http://www.squarefree.com/ bookmarklets/forms.html`	A javascript snippet that changes PSOT forms to GET (see *Protocol 1*)
Add searching to your site		
`ht://Dig`	`http://www.htdig.org/`	WWW indexing and searching system for a domain or intranet

that it suits you perfectly, and modifying a pre-existing set of links is so simple that it is well worth while. This modified/custom links page can be used to jump to all of your commonly used websites, perform automated searches of specific databases and more. It will work just as well stored on your computer's own drive as posted on a web server. Bookmarks stored on your web browser can also serve this function if properly organized, and this saves you going to the 'trouble' of learning how to make links in HTML. The name of each link is stored in your bookmarks directory. However, as links are simple to make and easy to update, and they can be published on the web if you choose to share your navigation tools with the world, this is the more useful option. If you choose to store all of your commonly used links as bookmarks, experiment with the 'Edit Bookmarks' feature of your browser to group related links in common directories (folders), and place subdirectories within directories to generate efficiently organized nested sets of bookmarks.

The simplest way to write a links page is save your bookmarks as an HTML file. Netscape stores its bookmarks in a specialized HTML format, so you can literally just put your 'bookmark.htm' file on a website, and it's ready to go. These bookmark files can be imported into Netscape, so you can use them as bookmarks, or you can just click on them from the web page. While Explorer does not natively store bookmarks this way, it can import and export to the Netscape format. This approach lets you use your browser's bookmark organizing features as a specialized editor just for links pages. Alternatively, you can write your own HTML using an HTML editor. Most modern word processor and web browser programs include an HTML editing function, and freeware and shareware HTML editors are also available via the web. One such editor, WebWriter, was used to write one of our websites. Creating a link using such software is simple, and a detailed description of the process is outlined elsewhere in this book (see Chapter 8). You can start from scratch or you can borrow a previously existing link set. This is done using the 'Save As' command while viewing a web page that is open within your browser. This command will open a panel asking you where to save the file and in what format, select 'Source' to save the page as HTML. Remember that material published on the web is copyrighted when it is created, so acknowledge the source of your starting links if you decide to publish your own modified version on the web. Once you have saved the source file open it from within your HTML editor using the 'Open' command. Delete links you do not like, cut and paste links to reorder them in a manner that better suits you, or add new links. Save your links page to your hard drive, then list it as your homepage using the 'preferences' settings of your browser.

3.1 Smart links

Links can be much more complex than simple web addresses to which your browser can jump. If you use your browser to request some sort of action from a website, for example performing a search, the instructions for that search are often contained within the information shown in the address window. For example, visit Yahoo at www.yahoo.com. Type 'Xenopus cytoskeleton' into the search window and click on 'search'. Your search directed your inquiry to a query program using your key words. If you look at the address displayed in the new address window it now contains your search terms. The key point is that if you now select the new *full* address and use it to create a link, the next time you use this link you will not need to type in 'Xenopus cytoskeleton', the link will return your search results automatically.

Some sites will not contain your search terms in the returned address window, including the many resources available through the NCBI such as PubMed; technically, this is probably due to the sites use of the 'POST' rather than the 'GET' method of receiving HTML form data (see *Protocol 1*). You can determine whether or not this is the case by looking for the presence of your search terms in the response window. However, even if the site you are using does not show your search terms in the returned address window it is still possible to build smart links containing your terms (author name, keyword, species etc). By building your own set of such smart links you can check on recent publications with a single click of the mouse button. The power of such links can go far beyond simple custom name or keyword searches by also including the filter functions available at the NCBI. For instance you can limit a PubMed search to specific time periods, so you can select for recent publications and exclude references that you reviewed the last time you checked. Detailed instructions for how to build such links is available from the NCBI (address below), but you can also experiment with the example links in *Table 3* to adapt them to your needs. The examples include simple links, links to an authors name, links to specific subjects, and more complex links with limits. If you are hesitant to go to the trouble of using an HTML editor simply use a text editor (such as SimpleText or NotePad) to open the document. Within the author link look for 'watson+jd' embedded in other information. Select just this part of the link and delete it, then replace it with your own surname and initials (e.g. I would type 'vize+pd'). You now have a PubMed link that will bring up all your own PubMed papers. You may wish to add this link to your own departmental website, and visitors will be able to search for all of your papers without you needing to keep updating your site because of your new productivity. Try changing the subject search link by switching 'Xenopus' to your own study organism, and 'microtubule+dynamics' for your favourite protein. Save the file, then open it in your web browser and click on it; this should return a PubMed search for your system and protein/gene.

Protocol 1

Creating 'Smart Links'

A 'smart link' to a database or search engine is one that has your query built into it, so that whenever the link is followed, the query is run against the database afresh.

Some web-based searches can be saved quite easily as smart links, others can be saved this way with a little manipulation, and some cannot be saved this way at all. The difference lies in how the query information is packaged to be sent to the database or search engine server.

The two common methods for sending queries from HTML forms are called 'GET' and 'POST'. In the GET method, the query information is attached to the end of the server's URL, with a question mark separating the regular URL from the query string.

A. Bookmarking GET method servers

Since servers that handle web requests using the GET method include the entire search strategy in the URL itself, such URLs can simply be bookmarked. Alternatively, copy the full URL from your browser's location window and paste it into an HTML link (see Chapter 8).

B. Servers that can handle either GET or POST method queries

If the server can handle requests using either method, you may be able to change the URL to include the GET-style query. PubMed falls into this category, as do most CGI servers using the Perl CGI.pm module (see Chapter 11). There may be problems if the form attempts to send more information than the GET method can handle (there is a size limit on URLs), but for straightforward queries, this approach frequently works. The hard part about doing this by hand is that you must know the names of the fields from the form, and special characters must be 'escaped' (e.g. spaces are replaced with plus signs etc.) It is often possible to edit the HTML for a web form to change its request method from GET to POST (and this can be a useful exercise for those wanting to learn HTML). In some cases a script can convert the request form to the GET method. *Table 2* lists a small script ('frmget') that modifies a request to create a GET method URL. It is simply a snippet of Javascript code that runs inside your browser to change a form's submission method. This Javascript program is small enough to be contained entirely in a URL itself, so you can simply add it to your list of bookmarks, and (try to) run it on any form.

Fortunately, we don't need to rely on technical tricks on the *PubMed site*, since it has a feature that creates GET method URLs for us:

1 Run your PubMed search as usual (see Chapter 2).

2 Click the 'Details' link below the 'Go' button near the top of the results page.

3 Click the 'URL' button below the detailed search box. This will create a GET-style URL, which will appear in your browser's 'Address' (or 'Location') window.

4 Bookmark this page, or copy it and paste it into a text file or HTML page. (Details on creating hyperlinks in HTML can be found in Chapter 8.)

C. Form translation service

The 'AutoPOST' program (see *Table 2*) lets you send it a request using GET, which it turns into a POST request. It runs your search for you and returns the results.

1 Enter the URL of the query form for your target database or search. AutoPOST will go fetch the form, and send you a 'mock-up' version of that form, which has been modified to talk to AutoPOST.

2 Fill out and submit the mock-up just as you would the original.

3 AutoPOST returns a GET-method URL that includes your search specifications.

4 Save (or bookmark) this URL, and click (or select) it whenever you want to re-run the search.

Table 3 Smart links in PubMed

Search for papers on Xenopus in the last 7 days
```
http://www.ncbi.nlm.nih.gov/entrez/query.fcgi?cmd=search&
term=Xenopus&db=PubMed&orig_db=PubMed&filters=on&pmfilter_EDatLimit
=7+Days
```

Search for papers by J. D. Watson
```
http://www.ncbi.nlm.nih.gov/entrez/query.fcgi?cmd=Search&db=PubMed&
term=watson+jd
```

Search for papers on the Xenopus heart that appeared in the last 120 days
```
http://www.ncbi.nlm.nih.gov/entrez/query.fcgi?cmd=search&
term=Xenopus+heart&db=PubMed&orig_db=PubMed&filters=on&
pmfilter_EDatLimit=120+Days
```

Search for papers on microtubule dynamics that appeared in last 120 days
```
http://www.ncbi.nlm.nih.gov/entrez/query.fcgi?cmd=search&
term=microtubule+dynamics&db=PubMed&orig_db=PubMed&filters=on&
pmfilter_EDatLimit=120+Days
```

---

4 How to contribute to the development of web resources

By far the greatest contribution most cell and developmental biologists can make to Internet resources in their area is to submit as much of their data as possible to the appropriate databases. A nice lab website may be a useful recruitment tool, but the information contained on such pages is generally of little use to the broader community. Databases depend on submitted data to ensure comprehensiveness, as no matter how well trained data curators are they cannot possibly understand your data as well as you do. By submitting your data to all relevant databases, and checking on the accuracy and quality of the data once it is added into the database, you are ensuring that your results will be accurately represented. It also ensures that the data is accurately referenced and that credit for the data has been correctly assigned. A number of database managers have indicated to me that they have never received any data submissions from the scientific community, despite their sites having many thousands (or tens of thousands) of visits per year.

The scientific community convinced the publishers of scientific journals long ago that it was essential for all published DNA sequence to be stored in centralized databases for the information to be useful. Most data from published experiments, not just protein and DNA sequences, could potentially be made publicly available in databases upon publication. Copyright laws interfere with this process to some extent, but even under the current restrictive system the data can be submitted to databases -the journals retain the rights to only the actual figures and text. Send other photographs and charts that illustrate the same data, or better yet, communicate with your major database and discuss how your raw data can be included at their site. Standards are now being adopted for interchange of gene array data (see *Table 4*). The availability of annotated primary data sets will at some point allow data analysis—not just data retrieval.

Table 4 Molecular biology and genomics resources

Blast	http://www.ncbi.nlm.nih.gov/blast/index.HTML	Compares sequences
eMotif	http://motif.stanford.edu/emotif	Protein sequence motifs
ESTs	http://www.ncbi.nlm.nih.gov/dbEST/	Expressed sequence tags database at the NCBI
Expression profiler	http://ep.ebi.ac.uk:80/	A clustering program for array analysis
Homologene	http://www.ncbi.nlm.nih.gov/HomoloGene/stats.HTML	Finds the closest related gene in other species
Unigene	http://www.ncbi.nlm.nih.gov/UniGene/index.HTML	Compilation of sequence data into individual contigs
Gene expression sites		
Mouse Genome Informatics	http://www.informatics.jax.org/	Genetic, genomic, and biological data on laboratory mice.
The Edinburgh Mouse Atlas Project	http://genex.hgu.mrc.ac.uk	3D atlas showing gene expression in mouse embryo
ZFIN: the Zebrafish Information Network	http://www.zfin.org	Zebrafish model organism database
Xenbase: a Xenopus web resource	http://xenbase.org	Cell and developmental biology of the frog, Xenopus.
Axeldb	http://www.dkfz-heidelberg.de/abt0135/axeldb.htm	A database focussing on gene expression data gathered by large-scale *in situ* hybridization screening in Xenopus laevis embryos.
Microarrays		
Micro Array Gene Expression Markup Language (MAGE-ML)	http://genomebiology.com/2002/3/9/research/0046	XML-based data format to facilitate the exchange of microarray data.
SMD	http://genome-www4.stanford.edu/MicroArray/SMD/	The Stanford microarray database. Stores raw and normalized array data
SMD Microarray Links : Software & Tools	http://genome-www5.stanford.edu/MicroArray/SMD/restech.HTML	Links to sources of various microarray analysis tools
ScanAlyze	http://rana.lbl.gov/EisenSoftware.htm	Microarray image analysis program
Cluster and TreeView	http://rana.lbl.gov/EisenSoftware.htm	Cluster analysis and visualization
XCluster	http://genome-www.stanford.edu/~sherlock/cluster.HTML	Cluster analysis on UNIX
ExpressionProfiler	http://www.ebi.ac.uk/microarray/ExpressionProfiler/ep.HTML	A set of tools for analysing and clustering gene expression and sequence data
Array Viewer	http://www.nhgri.nih.gov/DIR/LCG/15K/HTML/images.HTML	A Java Applet for analysing array data
ArrayExpress	http://www.ebi.ac.uk/arrayexpress/	Aimed at storing well annotated, structured microarray data using the MAGE-ML standard.
Gene Expression Omnibus (GEO)	http://www.ncbi.alm.nih.gov/geo/	Much less structured than ArrayExpress. This makes it easier to submit data, but harder to mine the data that has been collected.

5 Existing tools for investigating cell and developmental biology

Our fields of research can each be divided into two broad groups that are associated with distinct Internet resources—those that associate and order gene products with respect to cellular and developmental processes (molecular biology) and those that document the phenomenology of cell and embryo structure and function (biology). Molecular biology is served by a vast array of web resources for analyzing data (see Chapter 2). Biologists, on the other hand, are generally served by resources that catalogue and describe, rather than analyse. Sites serving the latter group (other than those mentioned above) can be tracked down using the approaches already described, while some of those serving the former group are discussed here. The peer-reviewed scientific literature remains the key reference resource to which all experimental data must ultimately be placed in context and thus serves the entire scientific community.

5.1 Literature

Various literature search services are available to help sift through the scientific literature. Such services evaluate the recently published literature and narrow down the deluge of data to a digestible level. As described previously it is simple to create your own custom 'smart links' that can serve a similar role by searching the PubMed database at your convenience (see Chapter 1). Custom links allow you to check on recent publications in your field with a single click, largely obviating the need for commercial services for routine literature analyses.

5.2 BLAST

Basic local alignment search tool (BLAST) plays a key role in the life of all molecular biology laboratories (see Chapter 2). This is the primary program used to compare sequences against the Genbank nucleotide database (1, 2). From finding homologs of a newly cloned gene to identifying plasmid contamination, BLAST is used by most laboratories on a daily basis. BLAST, or variations such as MEGABLAST, also provide the engines that drive many other informatics resources, such as UNIGENE and HOMOLOGENE.

5.3 Motif identification

Molecular biologists like to infer function from sequence. One way this can be done is by performing a BLAST search and finding a match to a gene encoding a protein with a know structure or function. But what if no match appears? Bioinformatics tools that can scan a sequence for similarity to functional domains also exist. An example site is E-motif at Stanford (*Table 4*). This web-based system allows you to cut and paste in your sequence then screen it against a structural motif database, helping you to identify potential functions.

5.4 EST databases

Expressed sequence tags (ESTs) are 'quick and dirty' single run sequences of cDNA clones. In theory a sufficiently large EST database will contain sequence for all genes expressed in the source library. The data is not meant to document the sequences of the expressed genes with complete accuracy—its purpose is to allow for identification of genes deserving further study. It is particularly useful for isolating orthologous genes. EST data is annotated to some extent, and each EST is compared with GenBank via BLAST and the closest match noted. If you would like to find the human version of a gene that looks interesting in a mouse (or even fly) study, simply search the human EST database (*Table 4*) for a match, then order the clone by phone.

5.5 Unigene

The unigene project at the NCBI (3–5) aims to group all sequences for each unique gene into a cluster, a unigene cluster. Any individual gene is therefore represented by a single cluster, rather than by a group of genomic, cDNA and EST sequences. This allows for a reduction in redundancy, since clusters, rather than individual sequences, will be used to select probes for building microarrays, orthologue searches, etc.

5.6 Homologene

Some of the best systems for the analysis of gene function are poorly suited to novel gene identification, and visa versa. Just as one can use genes isolated from good screening systems to pull out orthologous genes from good analysis systems using low stringency hybridization, one can use data on orthologous genes in tractable systems to extrapolate possible functions in intractable systems. For example, suppose you had isolated a human gene whose structure implicated it in embryonic patterning. For various ethical, legal, and technical reasons, the process of embryonic patterning is not easily studied in human cell culture. But if you could identify a previously isolated orthologue in zebrafish a vast amount of data on developmental expression patterns and even genetic mutant profiling may be instantly available. The availability of such information may allow you to jump a gulf in knowledge and design experiments suited to your own system.

Homologene is a tool for identifying potential orthologues. Pairwise searches of Unigene clusters in different species are used to classify the two most similar clusters in the two animals as 'calculated' orthologues. If additional data support the orthologue identification, these matched genes are referred to as 'curated' orthologues. The 14 organisms used in compiling Homologene (as of December 2002) include: four mammals, one amphibian, one fish, three invertebrates, and five plants (see *Table 4*).

The identification of orthologues can be skewed by the presence of different numbers of related genes in compared systems. If, for example, species A has one HOX gene, and species B has 10 HOX genes, the species A gene would be listed as the calculated orthologue of all ten species B genes. Despite this potential problem, over 7000 calculated orthologue identifications were consistent between more than one pair of species. With the rapidly expanding size of the EST databases for many organisms, Homologene promises to become an extremely useful tool.

5.7 Microarray analysis

Large-scale expression profiles of RNA samples can be generated using microarrays. The basic principle is to label an RNA sample and anneal the labelled sample to a grid of DNA sequences. The DNA sequences can be cDNA clones, exons, or oligonucleotides that correspond to expressed sequences. The binding of the labelled RNA is then scored, usually using digital imaging. The power of this technique is based on its scale, as robots can be used to array many tens of thousands of DNA spots on a single small slide or filter. This means that researchers can obtain very complex profiles and examine the expression of vast numbers of genes within a sample.

Microarray technology is a relatively new addition to the molecular tool kit, but database tools for analysing the masses of data being generated are rapidly becoming more sophisticated (6). Repositories of array data from model organisms used in developmental studies are available at the Stanford Microarray Database (7) (*Table 4*), as are various software packages for analysing such data. ScanAlyze is free software that can be used to read array output and translate the pattern of positive and negative spots into standard spreadsheets ready for data analysis. Cluster, or its UNIX flavour XCluster, can be used to analyse such spreadsheets and search for clustering-coordinated behaviours of different spots, as can ExpressionProfiler from the EBI. If different genes (spots) react in a similar manner, they may be regulated by similar factors and may also belong to the same or closely related pathways. This is potentially a powerful system to identify novel pathway components armed with nothing more than sequence data and labelled RNA samples.

6 Gene expression patterns

To understand how genes control developmental processes it is essential to know when and where each gene is expressed. Gene expression data is available on the web in a number of different formats. Perhaps the most simple are tables of RT–PCR data that score the presence or absence of gene transcripts in a wide range of embryonic tissues. More complex depictions are possible with photographs or images of embryos processed by *in situ* hybridization (*Table 1*). However, the data within the images cannot be used for searches or data analysis. For this reason, gene expression data is often annotated. In this context annotation refers to assigning descriptive words taken from a standardized vocabulary to image-based data. It is important to call different examples of the same things by the same words, and not use 'heart muscle' for one entry and 'cardiac muscle'

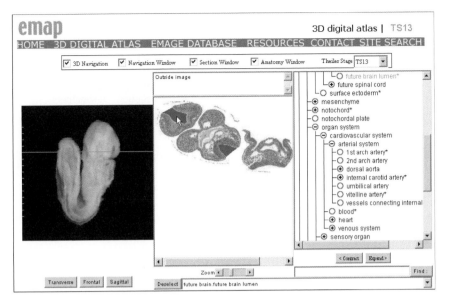

Figure 1 The 3D Atlas of mouse embryonic anatomy. This image was captured from the Edinburgh Mouse Atlas Project (EMAP) at `http://genex.hgu.mrc.ac.uk/Atlas`. The left side shows a 3D navigation frame, which lets you rotate the image of an embryo around the vertical axis. The horizontal line shows the level of the cross section to be displayed; it can be dragged to any desired level. The cross section in the centre is an image map; clicking a hotspot highlights the tissue, and its identity is shown in the tree of anatomical terms at the left of the figure.

in another. Good gene expression databases use carefully compiled annotation terms and standard nomenclatures that ensure consistency. More advanced controlled vocabularies re-organized into 'ontologies' in which similar or otherwise related words are cross-referenced (see *Table 1*). This allows automatic searching for synonyms, slightly more or less general terms, etc.

The most sophisticated *in situ* hybridization-based gene expression database currently available is the mouse atlas and gene expression database project in Edinburgh (*Table 1*). At this site 3D models of mouse embryos have been generated and gene expression patterns are linked with coordinates within these models. Tissue boundaries have been defined and users can colour code any desired set of embryonic structures (*Figure 1*). Systems such as this will hopefully soon allow us to actually analyse gene expression patterns and predict gene interactions based on both structure and co-expression.

7 Complex pathway representation

The signalling pathways that regulate cellular activity are extremely complex and difficult to represent or analyse using traditional media. The web provides an ideal tool for the display of such data, since hyperlinks allow you to jump from dimension to dimension to explore interconnectivity. Different transduction pathways

Table 5 Signalling pathway sites

Biocarta	`http://www.biocarta.com/`	Pathway information and discussion forums
Science's STKE	`http://stke.sciencemag.org/`	Collection of electronic networking tools for signal transduction researchers (registration required)
	`http://stke.sciencemag.org/cgi/cm/CMP_6634`	GαS pathway
	`http://stke.sciencemag.org/cgi/cm/CMP_6557`	PI3 kinase
	`http://stke.sciencemag.org/cgi/cm/CMP_6557`	wnt pathway
wnt Pathway	`http://www.stanford.edu/~rnusse/wntwindow.HTML`	
Homeobox genes	`http://copan.bioz.unibas.ch/homeo.HTML`	
MAPK pathway	`http://kinase.oci.utoronto.ca/signallingmap.HTML`	Image map of the main components of the mitogen-activated protein kinase pathway
Kinases	`http://www.sdsc.edu/Kinases/pk_home.HTML`	

often directly interact, so viewing a linear sequence of events is not an effective method to understand the cellular consequences of a signalling event. The signal transduction knowledge environment (STKE) is a project that aims to document connections and interconnections between transduction pathways. BioCarta has similar goals. Links to these and other websites displaying various transduction pathways are indicated in *Table 5*. Even with a site that only displays a linear pathway this medium is far more useful than print, as hyperlinks can lead users to related information and the site can constantly be modified as new information becomes available.

8 Summary and conclusions

All of the approaches discussed here are extremely powerful. Their power is derived from the ability to efficiently display and analyse complex biological data. As tools for linking these different databases are developed and we can link complex signal transduction pathways to array data, the analysis of cellular function *in silico* may be realizable. Until that time we have an extraordinary array of new tools at our disposal that can enhance the ways that we do science. Sequenced genomes, EST databases and tools such as homologene make it straightforward to identify candidate genes, robotics are automating the analysis of gene expression patterns, and tools for transfer of expression data to computers are becoming available. The fruits of genome-scale projects will be harvested on the Internet.

References

1. Altschul, S. F., Gish, W., Miller, W., Myers, E. W., and Lipman, D. J. (1990). Basic local alignment search tool. *J. Mol. Biol.*, **215**, 403–10.

2. Altschul, S. F., Madden, T. L., Schäffer, A. A., Zhang, J., Zhang, Z., Miller, W., *et al.* (1997). Gapped BLAST and PSI-BLAST: a new generation of protein database search programs. *Nucl. Acids Res.*, **25**, 3389–402.

3. Schuler, G. D. (1997). Pieces of the puzzle: expressed sequence tags and the catalog of human genes. *J. Mol. Med.*, **75**(10), 694–8. (Review.)

4. Schuler, G. D., *et al.* (1996). A gene map of the human genome. *Science*, **274**, 540–6.

5. Boguski, M. S. and Schuler, G. D. (1995). ESTablishing a human transcript map. *Nat. Genet.*, **10**, 369–71.

6. Gardiner-Garden, M. and Littlejohn, T. (2001). A Comparison of microarray databases. *Brief. Bioinfo.*, **2**, 143–58.

7. Sherlock, G., Hernandez-Boussard, T., Kasarskis, A., Binkley, G., Matese, J. C., Dwight, S. S., *et al.* (2001). The Stanford microarray database. *Nucleic Acids Res.*, **29**(1), 152–5.

Chapter 7
Internet collaboration

Chin Hoon Lau
Lagenda Knowledge Systems Sdn. Bhd., Johor Bahru, Malaysia.

David Atherton
Texas A&M University, Health Science Center, Cardiovascular Research Institute, Building 205, 1901 S. 1st St., Tempo, TX, USA 76504

Pinar Kondu
Iontek, Inc., Meridyen Is Merkezi Ali Riza Gurcan Cad., Cirpici Yolu No. 1/410, Merter 34010 Istanbul, Turkey

Robert E. Gore-Langton
The EMMES Corporation, Rockville, MD, USA.
e-mail: robg@internetbiologists.org

Zev Leifer
New York College of Podiatric Medicine, New York, NY, USA.

1 Introduction

1.1 Overview

This chapter surveys the types of technologies, applications, and organizational approaches that best support the needs of scientific collaboration. Although briefly covering the foundations of Internet communications technologies, such as e-mail and newsgroups, emphasis has been placed on solutions to support real-time interactions and shared tools since we believe these provide the greatest opportunities for meaningful collaboration. Some of these tools also provide the greatest technological challenges, but the rewards in terms of creativity and efficiency are often worth the effort.

We provide Uniform Resource Locators (URLs) for the key sites to jump-start your use of the Internet in collaboration. A companion web page provides hyperlinks to these sites and to other useful resources (see *Table 1*). We invite readers to share new links and resources concerning internet collaborations (see contact information on the web site).

Table 1 Overview and basic tools

Overview		
Resources for Internet collaboration	A companion web page to internet collaboration, Chapter 7, In: *The Internet for molecular biologists*, Oxford University Press, Oxford, 2004	`http://www.internetbiologists.org/collaboration/`
Challenges of Internet collaboration		
Schrage, M. Concepts of collaboration	Shared Spaces as Collaborative Tools. Information Technology 'summIT'. (Internet Archive March 10, 2000)	`http://web.archive.org/web/20000310033144/www.dtic.mil/summit/ma01.html`
Schrage M. Leading lights: Technology Designer Michael Schrage	Interview	`http://www.webcom.com/quantera/schrage.html`
Asynchronous communications and E-mail		
Multipurpose internet mail extensions (MIME)	MIME extends the format of internet mail to allow non-US-ASCII textual messages, non-textual messages, multipart message bodies, and non-US-ASCII information in message headers	`http://www.oac.uci.edu/indiv/ehood/MIME/MIME.html`
Eudora version 5.2	E-mail client from Qualcomm	`http://www.eudora.com/`
Wincode Encoder/Decoder	Freeware utility supporting UUencode, BASE64, BinHex and other formats for Windows 3.x and 95	`http://www.winsite.com/` Enter search for wincode
In a Nutshell. Using wincode	Unofficial tips (and alternate download for wincode)	`http://www.telecommunications.com/nutshell/wincode.htm`
Pretty Good Privacy (PGP) Freeware	MIT Distribution Center for PGP cryptographic software	`http://web.mit.edu/network/pgp.html`
Mail Lists		
LISTSERV	Commercial mailing list management software (list server)	`http://listserv.acsu.buffalo.edu/`
Majordomo	Another widely used list server favoured by Internet Biologists (IB)	`http://www.greatcircle.com/majordomo/`

Name	Description	URL
CataList reference site	Official catalogue of LISTSERV lists (75,353 lists available)	`http://www.lsoft.com/catalist.html`
Lyris.net	E-mail marketing resource guide. Link to ListManager list server	`http://www.lyris.net/`
QuickTopic.com	Create a hosted single topic discussion list (bulletin board)	`http://www.quicktopic.com`
QuickTopic Document Review	A private space for consolidating comments from collaborators on an HTML or Word document under review	`http://www.quicktopic.com/cgi-bin/docreviewintro.cgi`
Asynchronous communications and E-mail	Program for Internet News & Email (PINE)	`http://www.washington.edu/pine/`
	An e-mail program with a full-screen terminal interface developed at the University of Washington. There are both UNIX and PC versions.	
Topica	Manage and participate in e-mail discussion lists (free membership registration). There are currently 49 biology discussion groups. Topica acquired Liszt Newsgroups but does not support newsgroups	`http://www.topica.com`
Newsgroups and Bionet		
GoogleGroups	Search and browse Usenet discussion forums (formerly DejaNews)	`http://groups.google.com/`
BioSci	Set of electronic communications forums (bionet newsgroups and parallel e-mail lists) used by biological scientists worldwide	`http://www.bio.net/`
BioSci bionet Newsgroup Archives	Advertising on this site supports newsgroup maintenance	`http://www.bio.net/archives.html`
Marcato, D. Host a Discussion Forum with the Windows 2000 NNTP Service	News server software built into Windows 2000. May be installed at time of operating system installation or at a later time	`http://www.microsoft.com/mind/0100/NNTP/NNTP.asp`

1.2 Challenges of Internet collaboration

Successful collaboration on the Internet, in science or any other discipline, requires familiarity with the enabling technologies, as well as the social and behavioural aspects governing human intellectual interactions. While technical publications tend to focus on the first of these requirements, true collaboration via the Internet depends to an even greater extent on the human challenges of communication in often cross-cultural and interdisciplinary circumstances.

The objective of any collaboration should be to achieve creative results that are more than simply the sum of the individual contributions (see Schrage, *Table 1*). At best this requires a shared workspace, well-defined objectives, an understanding of the special knowledge and skills that each collaborator will contribute, and the ability to motivate, organize, and facilitate a team effort. These are challenging requirements even in a face-to-face setting, but are more difficult with any electronic communication that lacks both the immediacy of voice and the many nuances of human communication (e.g. facial expressions and body language) that we normally depend on. If available communication tools are inappropriate or technical know-how inadequate, that only adds to the barriers and can be a frustrating experience. Fortunately, there are now many resources available to support the technical aspects of Internet communications, as well as to provide practical training in Internet use. Some of these resources focus on the needs of the bioscientist.

1.3 Basics of Internet communications

Collaboration via the Internet is enabled by one or more modes of communication—text, images, voice, or video. In addition, a number of technologies permit sharing of common resources including files, calendars, custom applications, web pages, archived messages, etc. The success of these means of communication, and ultimately the collaboration, depends on the ease with which the tools can be implemented and the clarity of understanding achieved between the collaborators.

The choice of communications tools often presents a dilemma in balancing ease of use with richness of features, from the simple and more robust (e.g. e-mail, mail lists and text-based web-enabled platforms) to the more sophisticated and generally more complex (e.g. Multi-User Virtual Environments and systems with audio/video capabilities). The correct balance for any collaborative venture depends on many factors, including the minimum technical capabilities essential for the particular collaboration, the individual and collective Internet experience of the collaborators, the bandwidth and reliability of the Internet connections, cross-platform limitations, and the availability of any third-party expert advice and support.

In general, it is recommended to begin with the simplest modes of communication that meet the collaborative need, and to add new capabilities as needs expand. Upgrading your Internet communications to maintain an efficient collaboration and to provide the greatest compatibility for current and new collaborators is an ongoing process. Average processor speed, memory, and bandwidth capabilities are steadily on the rise and show no sign of slowing. Internet tools, often driven by e-business forces rather than scientific requirements, evolve rapidly to take advantage of these changes. You will need to remain current on these developments and respond appropriately to stay at the forefront of Internet communications. The evolving communications capabilities of your collaborators, and especially the minimal standards to be supported in your collaboration need to be considered.

2 Collaborative environments and shared tools

2.1 Asynchronous communications

Communications without simultaneous interaction or where significant delay occurs between consecutive messages are considered asynchronous. E-mail, mail lists, and newsgroups are asynchronous tools.

2.1.1 E-mail

E-mail is the most widely used Internet technology and will be the starting point for most Internet-based collaborations. E-mail has the advantages of being simple to set up and use, relatively fast at transmitting text messages, and generally private although not secure.

Depending on the mail client used, e-mail can be enhanced by enabling HTML to display URLs as hyperlinks, and to send/receive a variety of binary file attachments (documents, images, multi-media presentations, programs). E-mail also works across all computer platforms and is not unduly limited by the speed of Internet connections or bandwidth.

E-mail comes in several flavours, including TCP/IP protocols for incoming and outgoing mail (Post Office Protocol, POP3, and Simple Mail Transfer, SMTP, respectively), shell accounts (with UNIX), and web-based mail accessed via your web browser. These mail systems have significant technical differences and different user interfaces, but their functions are similar and will not be distinguished here. For Windows PC or Mac operating systems there are a large number of E-mail client programs with better or worse interfaces, but again basic functions are similar. Unix users will likely use Pine, a command-based program accessed on the server by a terminal program (see *Table 1*). AOL subscribers are limited to the provided mail program.

For serious e-mail collaboration one of the standard mail clients with the following minimum features is recommended: Multipurpose Internet Mail Extensions (MIME)-compliant encoding for file attachments, comprehensive address book, and message filtering into mailboxes. These features are standard on most popular mail clients, such as those from Microsoft and Netscape, as well as higher end separate clients such as Eudora (see *Table 1*). See *Protocol 1* for suggestions on how to use e-mail efficiently for collaboration.

Protocol 1

How to use e-mail efficiently for collaboration

1 Use (or install) a full-featured MIME-compliant mail client. This will in most cases permit safely sending binary file attachments. Files sent with UUcoding (DOS or UNIX) or BinHex (Mac) may also need the WinCode utility program for Windows 3.x and 95 for unencoding (see *Table 1*).

2 Track your collaborators and their contact information in the address book; assign nicknames to speed sending messages.

3 Filter mail (incoming and/or outgoing) into separate mailboxes (folders) according to topic, project, or collaborator. You may alternatively just filter a copy into folders for message archiving.

4 Do not send files in proprietary formats or with software versions that are not available to all collaborators; in most cases the recipients need the same software to open or run a file as was used to create it.

5 Use local mail distribution lists, supported by some mail clients, for simple or temporary group mailings, but consider list server or other web-based mail lists where there are many collaborators or a lengthy collaboration.

6 If the collaboration involves particularly sensitive or trade secret subject matter, use a strong encryption such as PGP (see *Table 1*).

2.1.2 Mail lists and discussion spaces

Some would consider mail lists (also known as list servers) to be the single most valuable tool for global online collaborations. Mail lists have all the advantages (and some of the disadvantages) of e-mail and then some. As in e-mail, messages are received and sent via your normal e-mail service and mail client. However, instead of addressing messages to each of your collaborators individually, you simply address messages to the mail list name. The mail list server then distributes messages to all of the list subscribers, and in a more efficient manner than e-mail.

The main advantage of mail lists is the elimination of the need for individual subscribers to locally maintain distribution lists on their mail client, and to ensure proper distribution to all members of your collaboration. However, the responsibility for administrative overhead then shifts to the list administrator, who may choose to manually approve subscriptions and will need to resolve list problems. Much of the overhead can be handled by automatic subscription/unsubscription, but this relies on individual members of your collaboration to subscribe themselves to the list. List servers can have the additional advantages of pooling multiple messages from multiple sources into threaded message archives.

Two of the most widely used list server programs (see *Table 1*) are: LISTSERV (the first list server) and Majordomo (the one used by IB). *Protocol 2* describes how to subscribe to a Majordomo list. Although other list servers are generally similar, commands, syntax, and some other features vary.

Protocol 2

Subscribing to a Majordomo list server

1 Address your e-mail message to the list host in the exact manner indicated in the notice or invitation to join. For example, majordomo@host_name, where host_name usually consists of server_name.host_computer.highest_level_domain; each part of the host name must be completed according to the actual host name for the list. It is important to remember that subscriptions are not sent to the mail list itself.

2 In the body of the message type: subscribe list_name, where list_name is the actual name of the mail list.

3 Do not fill out the subject line.

4 Send the message and wait for an e-mail confirmation. You may then post your first message to the list mail address (e.g. send to: list_name@host_name).

5 Various information concerning the list can be obtained (if you are curious) by sending majordomo special commands (e.g. sending to majordomo@host_name, with message 'help list_name', where the host_name and list_name refer to the actual host and list names, will return a list of possible commands and instructions.

6 List administrators or other approved persons can use a password to access and make changes in the subscriptions and software settings.

7 If created separately, web-based HTML forms may also be used to make individual subscriptions and to administer lists.

Joining some mail lists may involve simply filling out a web-based form or check box. Examples are discussed in Section 2.2.2 on Integrated web sites. These may use proprietary systems, so follow the instructions.

There are many possible list servers with open subscription lists; for example, see the CataList directory for LISTSERV lists (see *Table 1*). For purposes of research collaboration you may seek out lists focused in your own discipline, perhaps maintained by professional societies or other scientific organizations. Mail lists are also often used for one-way distribution of information concerning seminars, conferences, current contents of journals, etc., which may in themselves assist in finding contacts for collaboration.

As an alternative to the traditional mail list, commercial list manager software developed for e-mail marketing may also provide the features you need. For example, Lyris.net for opt-in mail is suitable for newsletters and announcements (free for up to 200 members).

A free hosted discussion space such as QuickTopic is a web/e-mail hybrid that provides a very easy route to creating a single topic discussion group. QuickTopic Document Review, either hosted or installed on your server, allows comments on any html or Word document to be created and viewed by the members in a private space (see *Table 1*).

Topica, which acquired the Liszt of Newsgroups in 1999, now provides hosted management and participation in e-mail lists and discussion groups. Also see Section 2.2.2 for integrated web sites with similar hosted services.

2.1.3 Newsgroups and Bionet

Newsgroups (principally USENET) provide another route to Internet collaboration (see *Table 1*). How effective this might be will largely depend on how effectively you use this medium to identify possible collaborators and make yourself known to others. Bionet, one of the 12 main newsgroup hierarchies, can be an incredibly valuable source of biological information provided by (in many cases) experts like you, and a potential source for collaboration. At last unofficial count, Bionet had about 119 newsgroups on distinct topics, with 27 groups concerning different areas of molecular biology; the actual number of newsgroups may depend on which groups your Internet service provider supports.

Despite this wealth of opportunity, several factors detract from the value of newsgroups:

(a) Newsgroups can consume an inordinate amount of your productive research time.

(b) Newsgroups are prone to distractions such as 'spamming', the posting of unsolicited and/or unrelated advertisements to multiple newsgroups, and 'flame wars', hostile and provocative messages that do not usefully contribute to the subject being discussed.

(c) To protect their own privacy and prevent receiving spam via e-mail, users may post messages anonymously. This may make it difficult to contact these users other than through the newsgroup.

How do newsgroups work? Each newsgroup provides discussion 'threads', in which an original article and each of its responses are linked together. Articles are read and responses posted using a news reader that communicates using Network News Transfer Protocol (NNTP) with a local server (known as a news daemon) provided by your Internet service provider. Simple news readers are often built-in with your browser and mail client, or you can download shareware or free news readers that may have additional features. No formal subscription to newsgroups is necessary, although subscribing through your reader will then display only the groups of interest. Alternatively, newsgroups may be located on a website. BioSci is the official bionet archive site; advertising on the site supports newsgroup maintenance (see *Table 1*). Current and past discussions are archived. You may respond to articles directly on the newsgroup, or to individual authors if an e-mail address is provided. Posted messages are generally accessible for 1–2 weeks before being erased. To facilitate the efficient use of newsgroups in your research and collaborations, you may be able to set filters and automate the process of article retrieval using the reader software.

Archived newsgroup messages may also be found in online directories at GoogleGroups (which acquired DejaNews) (see *Table 1*). Information on the newsgroups to which specific authors post articles is also available.

Creating your own newsgroup/discussion forum on your server is also possible using the NNTP service built into Microsoft Windows 2000 or the NNTP Option Pack for NT (see *Table 1*).

2.2 Synchronous communications and shared tools

Synchronous or real-time communication is usually superior to e-mail when the purposes include brainstorming new ideas, reaching a group consensus on complex issues, or simply meeting new collaborators to explore future research directions.

Some of the most useful tools for synchronous communications come integrated with a wealth of other shared tools and resources. Web-based environments can provide complete platforms to manage administrative aspects of your collaborative ventures. Multi-user virtual environments may provide access to whole research or educational communities, with shared spaces and resources, sophisticated communications, and virtual devices that mimic the real world. These two quite distinct collaborative environments are often used in a complementary fashion.

2.2.1 Internet chatrooms, instant messaging, and whiteboards

The ubiquitous chatrooms of social entertainment (e.g. Internet Relay Chat or IRC, 2D/3D chat formats, and many web-based chats) generally lack the professional focus and privacy needed for scientific collaboration. However, it is expected that more academic and science web portals will in the future offer shared synchronous communication on their sites. Free or low cost chat-hosting services and chat servers are also available for those wishing to set up their own chatrooms (search for current examples of these products on the ZDnet site; see *Table 2*).

Instant messaging (IM), with popular messengers such as ICQ, Yahoo! Messenger, AOL Instant Messenger (AIM) and MSN/Windows Messenger, may also support collaboration once contact is established. While these tools greatly accelerate communications their value for scientific collaboration is limited by poor interoperability and security concerns. IM clients such as Trillium Pro and the IM protocols offered by Jabber have to some extent addressed the interoperability problem for connecting to legacy IM systems (see *Table 2*).

The Groupboard server (see *Table 2*) provides free or full versions of its collaborative whiteboard, which also includes simple chat and message boards; these have been installed on laboratory home pages or for use in courses.

Table 2 Specialized tools

Internet chatrooms, instant messaging and whiteboards

ZDNet	A convenient source of software reviews and downloads for Internet tools	http://www.zdnet.com/
Groupboard.com	Install a collaborative whiteboard, chat or message board on your web page. Demo available	http://www.groupboard.com
ICQ homepage	Acronym intended to sound like 'I Seek You'. Prototypic instant messenger software with many features. Allows private messaging, but like most IM systems the Interest Groups are not professionally oriented	http://web.icq.com
Yahoo! Messenger	Instant messenger from the Yahoo family of Internet services	http://messenger.yahoo.com
AIM	A widely used instant messenger for AOL customers	http://www.aim.com
MSN Messenger for Windows	Another instant messenger from Microsoft	http://messenger.microsoft.com
Cerulean Studios Trillian Pro	An instant messenger client intended to provide interoperability with other systems	http://www.ceruleanstudios.com/trillian/index.html
Jabber Software Foundation	An instant messenger protocol designed to provide interoperability	http://www.jabber.org

Integrated websites

Yahoo! Groups	Everything in one site for mail lists, file sharing, chat, etc., was called eGroups before becoming part of the Yahoo family of services. A favorite of IB.	http://www.groups.yahoo.com
Intranets.com	A commercial all-in-one collaboration site with many features	http://www.intranets.com

112

Blackboard.com	Develop a free online course with many features. Commercial versions of this courseware are widely used by universities and colleges. (There are many other courseware products and hosted services)	http://company.blackboard.com/courses/index.htm

Multi-user virtual environments

LambdaMOO	The prototypic Multi-user Object Oriented (MOO) collaborative environment	http://trace.ntu.ac.uk/community/Lambda.htm
MOO home page	Links to MOO resources	http://www.moo.mud.org/
LambdaMOO's Programmer Manual (April 1996)	Resources for building in a LambdaMOO-based environment	http://www.bvu.edu/ctown/Progman/ProgrammersManual_Toc.html
WebDeveloper.com. VRML	Virtual Reality Modeling Language (VRML) resource	http://www.webdeveloper.com/vrml/
BioMOO home page (now defunct—see web archive)	Past example of a premier biology community online (Internet Archive February 24, 1999). BioMOO was the collaborative home of IB and many other biology collaborations until 2002	http://web.archive.org/web/19990224020349/ http://bioinformatics.weizmann.ac.il/biomoo/
BioMOO: Teaching Biology in Virtual Reality by Clare Sansom	An article on teaching in a MOO environment, based on the now defunct BioMOO	http://science.uniserve.edu.au/mirror/vCUBE97/html/clare_sansom.html
Diversity University (DU) at DU Educational Technology Services, Inc. (non-profit educational organization)	A collaborative MOO currently used by IB. DU hosted IB online lectures and discussions for its 'Perl for Biologists' course in 2002–2003	http://www.du.org

Table 2 (*Continued*)

IB MOO Guide for DU by Zev Leifer	Describes the options for connecting to DU and accessing the IB classroom	`http://www.ib-perl.org/class/MOO/MOOguide.html`
Collaborative Virtual Workspace (CVW)	Find information on the open source CVW prototype collaborative environment. This 'pre-built' MOO environment based on LambdaMOO was developed by Mitre Corporation and released to the public domain	`http://cvw.sourceforge.net/cvw/info/CVWoverview.php3`
SourceForge	Official site for open source software	`http://sourceforge.net`
SjSWorld Community Forum	Instant Messaging simply explained	`http://pub169.ezboard.com/fsjsworldforumsfrm7.showMessage?topicID=1.topic`
Other real-time tools		
Conferencing on the Web by David R. Wooley	Categorized listing of software for Internet discussion and collaboration	`http://thinkofit.com/webconf/index.htm`
ConferZone	Resource Center, Newsletter, and Vendor Information	`http://www.conferzone.com/`
CollabTools	Listing of tools for Internet collaboration	`http://www.voght.com/cgi-bin/pywiki?CollabTools`
Microsoft Windows Technologies Windows NetMeeting	Windows software for online collaborative meetings. Supports A/V conferencing, whiteboard, chat, and file sharing. Comes with Windows 2000	`http://www.microsoft.com/windows/NetMeeting/Features/default.ASP`
Net2phone	One of many products for Internet telephony (low cost service)	`http://www.net2phone.com/`
pcAnywhere	Symantec's remote control software. May be useful in controlling collaborative applications on another computer and for file transfers	`http://enterprisesecurity.symantec.com/products/products.cfm?productID=2`

Title	Description	URL
Community of Science (COS)	Various tools to assist professional scientists and societies. Access to databases of funding information, experts, and online reference databases through the COS Workbench	http://www.cos.com/
COS Abstract Management System	COS Abstract Management System is suitable for managing review of abstracts for scientific societies. A demonstration is available	http://ams.cos.com/demo/overview.shtml
ePanel from Humanitas. The On-Line Panel Evaluation Website.	Software for collaborative panel evaluations. In use by government agencies. Supports Windows and Macintosh	http://www.epanel.cc/
NIH VideoCasting at the NIH Center for Information Technology (CIT)	What is NIH VideoCasting	http://videocast.nih.gov/videocastinfo.asp
NIH VideoCasting	Direct access to NIH current events and archived scientific lectures on video	http://videocast.nih.gov/
Usability First. Your online guide to usability resources	Link to groupware resources	http://www.usabilityfirst.com/
Udell, John. Internet Groupware for Scientific Collaboration	In depth guide to groupware for scientific collaboration (ca 2000)	http://udell.roninhouse.com/GroupwareReport.html

2.2.2 Integrated web sites

A number of highly integrated sites offer free collaborative web-based facilities including asynchronous and synchronous communications, as well as shared resources for groups (see *Table 2*). Examples are 'Yahoo! Groups' (formerly eGroups), Intranets, and, for instructional courses, Blackboard. Typical features include password-protected secure access, chat with separated text input and output, mail lists, file uploading, hyperlinks, simple databases, calendars, and simple member polls. Blackboard also provides discipline-specific resource centres and course management tools. Offering an online course or workshop as part of your strategy to build collaborations is now relatively easy.

However, web-based collaborative sites have limitations. Notably, some sites do not fully support all web browsers, recording/archiving of text discussions may not be possible, connection times may be slow when these popular servers are busy, and there is no guarantee that current policy or services are permanent (see individual sites for pertinent details). Principal attractions are simplicity of getting started (taking only a few minutes), shallow learning curve, ease of managing multiple groups and usually excellent online help.

2.2.3 Multi-user virtual environments

2.2.3.1 *Introduction to MUVEs, MUDs, and MOOs*

Multi-User Virtual Environments (MUVEs) are shared spaces based on a Multi-User Dimension or Domain (MUD). MUDs are programs that reside on a server to which multiple users can concurrently connect over the Internet via telnet or specialized MUD clients. MUDs provide a shared, text-based, virtual environment in which participants may 'travel', explore, interact, and communicate. Although MUDs originated in the role-playing world, applications have evolved into sophisticated educational and research communities in the form of object-oriented MUDs called MOOs (MUD, Object-Oriented).

LambdaMOO, which first operated in 1990, is the prototypic MOO. It was set up as a 'social' MOO. The next important step was the development of BioMOO, specifically designed as a meeting place for bioscience researchers and teachers. Many of the projects described in Section 3 took place in BioMOO (a description of this now defunct site can be found in the Internet Archive; see *Table 2*). That venue was discontinued in 2002. Courses taught by IB have moved to Diversity University (DU), which will be the primary area for current MOO activity, as described below (see references in *Table 2*).

The MOO language treats everything as an object (including users) and allows simple or complex objects to be created, viewed, and manipulated in virtual space, while the multi-user capabilities allow members to simultaneously share these experiences. Teleporting from room to room, picking up and examining objects, paging members, sending MOO mail, checking out a member's research interests, and reading a poster presentation are everyday events in an environment such as DU, an online meeting place in the form of a virtual university campus.

MOO 'wizards' and voluntary programmers do most of the programming, so the MOO land is already well endowed with useful resources even for the newcomer. However, creating and customizing your own virtual office is part of the fun of DU and can usually be achieved after several days' experience.

2.2.3.2 *MOO communication and navigation*

Synchronous group communication is still one of the primary functions of MOOs and requires some explanation. Like web-based chats, text messages are sent to the shared window in real-time. However, that is where the similarities end. Communication in a MOO is not web-based, although registration and connection are often initiated using a web browser. Instead text messages are entered either via a telnet client (usually provided with your operating system) or using a specialized MUD client, which must be downloaded and installed. Both clients usually provide superior performance to a web-based chat application. MUD client interfaces vary, but often feature automatic host connection, separated input/output text, not so useful push-button 'emoticons' (i.e. to display textual symbols expressing your feelings), and extremely useful logging of displayed text to a file.

MOO communication, navigation, and other activities are for the most part controlled by text commands. Correct syntax is important. Thus, for user RobG in DU to send a greeting to the shared telnet window, one would enter <say Hello> or shortened to <"Hello> to have displayed 'RobG says Hello'; do not type the brackets, which are used here only to distinguish the entered commands. There are many commands possible, and those needed to start in DU are provided in the MOO Guide (see *Table 2*). Frequently used commands for communications in DU permit paging, whispering (for private discussion), talking on intercom channels, and pasting multiple lines of text to the display. To move the user through the MOO's space and rooms, other commands display a map, 'move' you in a direction, or teleport you to a specific location. The <who> command shows who is logged on and their location, while the <join username> command will teleport you to the same room as your course instructor or colleague (hopefully expecting you).

Functional objects in DU are picked up and examined by other commands. Examples of everyday objects useful in collaborations on DU include signs, notes, notice boards, notebooks, recorders, slide projectors, lectures, and tutorials. Other more sophisticated shared research tools are discussed later with collaborative applications in Section 3.

So far we have only described a text-based telnet or client window into the MOO. However, MOOs can also be visualized, navigated, and explored using a web browser; the web window can be either integrated to 'follow' the telnet window in virtual space or be independent. The web window provides one of the most useful and more visually appealing ways to examine shared resources in the MOO, using a mouse or pointing device to select objects of interest. Most of DU's web pages also display a map of DU in an image map form that can hyperlink

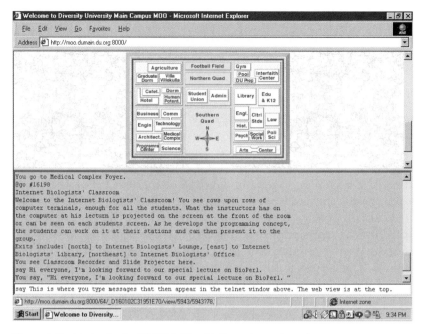

Figure 1 Navigating in the DU MOO. The 'bird's eye' overview of the virtual campus is shown in the map at the top, while navigation options for moving from room to room in a 'building' are given verbally in the text box below. (See Section 2.2.3.2.)

directly to specific locations (*Figure 1*). The web view may be enhanced with VRML to show objects as 3-dimensional representations.

2.2.3.3 *Visiting or starting a MOO*

If your appetite is now whetted, the next step is to visit a MOO in person. Because of the potential to combine a telnet or MUD client window with either an integrated or independent web window, there are several variations on how to establish a connection to DU. Step-by-step protocols for each configuration are provided in the MOO Guide (see *Table 2*). You may also visit DU as a guest by selecting the Java-based combined telnet/web interface option at the DU Web Gateway page; only a Java-enabled browser is required.

Individuals or organizations interested in developing their own educational MOOs are referred to a collection of essays on the subject (1). You may also wish to consider the Collaborative Virtual Workspace (CVW, version 4.0) developed by Mitre Corporation and now released to the public domain. Although the LambdaMOO source code on which CVW is based contains very little in the way of a virtual environment, the CVW MOO server contains a complete pre-built environment, and is extensible for multicast videoconferencing. Further information is available at SourceForge (see *Table 2*).

Protocol 3

Using shared tools to run a research collaboration

1 Set up your collaboration using e-mail.

2 Start a group/intranet web site (e.g. Yahoo! Groups or Intranets) to manage your routine group communications (chat and e-mail) and file sharing. A list server program provides a good alternative if a mail list is the only need.

3 Join an online community such as DU and build a virtual office where you can meet with your collaborators. Create virtual objects from pre-built generic objects to support your collaboration (e.g. notes, recorder, slide projector, etc.). The advanced MOO user might create new functional tools to be shared.

4 Join a working group to help develop an existing virtual laboratory, or set out to develop a new laboratory project.

5 Create a laboratory web page to share your research interests, publications, and ideas with your collaborators; customize it with a whiteboard or message board.

6 Keep aware of new communications and collaboration technologies and explore opportunities that will improve your distance collaborations.

2.2.4 Other real-time tools

An ever-changing selection of freeware, shareware, commercial software and hosted services may also find a place in your collaborative toolkit. However, suitability for large or global collaborations may be limited by the need for all participants to install the same specialized software, incomplete cross-platform support, and often-poor interoperability of competing products. Also, hardware and bandwidth requirements may limit accessibility for high bandwidth applications such as videoconferencing. Comprehensive listings of software and services for real-time conferencing, forums, and collaborative work environments may be found at Conferencing on the Web, ConferZone, and Collaborative Tools (see *Table 2*). There is no shortage of possible solutions for online collaboration, but identifying the product to match your every need and budget can be challenging.

Several categories and a few typical examples of proprietary tools (see Other real-time tools, *Table 2*) are: audio and videoconferencing, for example, NetMeeting; Internet telephony or Voice over Internet (VoIP), for example, Net2Phone, and many others; and remote viewing or control, for example, pcAnywhere. Microsoft NetMeeting with video, chat, whiteboard, program sharing, and desktop sharing, is the closest thing to a collaborative environment, but only the Windows operating system and Internet Explorer 4.01 or later browsers are supported. NetMeeting comes already installed in Windows 2000. You are strongly advised to conduct your own searches for products that specifically meet your needs.

The Community of Science, a site for the global R&D community, provides professional societies with web-based tools for conference administration, online authoring, and peer review. Its Abstract Management System can be integrated into a society's own website to permit society members to collaborate online in the creation, editing, and submission of scientific abstracts.

For institutional or corporate online evaluation of proposals (e.g. grant applications, clinical investigations, etc.) there is ePanel, a web-based real-time collaborative environment that facilitates and manages the evaluation process. The virtual conference room of ePanel allows users at remote locations to collaborate on creating a written evaluation summary, while using additional tools for scoring, voting, and speaker queuing, in combination with conference calls.

Videocasting, although not used for synchronous communications and not shared between collaborators, is included in this section because it may be one of the most useful and readily accessible tools for awareness of current research at distant locations. Videocasting sends digitally encoded streaming video and audio data from a server to clients, where a media player such as RealPlayer decodes it. Lectures in real-time may be viewed by many individuals concurrently using an efficient unidirectional multicasting. Bi-directional unicasting, although less efficient, can provide archived video on demand. As an example, the US National Institutes of Health lists videocast lectures in real-time, as well as lectures archived on their site.

The foregoing sections on shared environments and tools have necessarily been incomplete, primarily focusing on tools most familiar or most likely to be useful to the research scientist. If you wish to follow up with further sources on groupware, the term used to describe all forms of electronic technology for person-to-person collaboration, you may refer to the book by Coleman (2) or go to Usability First's site and follow the links to Groupware. The 'Group Links' page on that site provides many resources concerning groupware and the related field of computer-supported cooperative work (CSCW). John Udell's article provides an in-depth review of groupware issues and products (ca 2000) for scientific collaboration, including group discussion, broadcasting and monitoring news, and scientific publication (see *Table 2*).

Protocol 3 provides suggestions on how to use some of these shared tools to run a research collaboration.

3 Collaborative applications in the biosciences

Having described the various tools available for collaboration via the Internet, the next step is to see how these tools have been used. The ultimate goal is to use these examples as models and apply them to new and personalized projects. The examples fall into four categories: (1) Courses, (2) Conferences, (3) Laboratories, and (4) Research and Project Collaborations.

3.1 Courses

There is an ongoing need for education in the biosciences, whether it is in providing 'old' knowledge to new audiences or new findings to experienced professionals. Very often the opportunity to attend courses or conferences or to collaborate is limited by geography, politics, funding, or the difficulty in traveling to off-site locations because of commitments or physical disabilities. Online attendance via the Internet is an ideal solution to many of these problems.

A number of online courses have been successfully run during recent years, utilizing a combination of asynchronous (e-mail) and synchronous (MOO) methodologies.

IB (see *Table 3*) is a group dedicated to promoting the use of the Internet in bioscience research. They have produced three courses, 'Bioscience Resources on the Internet (BRI)', 'Doing Biological Research on the Internet (DBRI)' and 'Perl for Biologists'; BRI and DBRI have had multiple international editions while the Perl course completed its first edition in 2003. The faculty was international (all of whom knew each other only via the Internet; see below). The student body (about 20 graduate students and working professionals) were recruited, screened, and accepted online. The courses were given over a period of 3–4 weeks (3 months for the Perl course). The BRI course covered techniques of online collaboration and the use of tools. The advanced course (DBRI) covered such specific applications as multiple sequence alignment, virtual reality in biology, online laboratories and molecular visualization. The Perl course taught a computer language much used in bioinformatics (see Chapter 11). Assignments were given, responses were made available to all online and the group discussed the results. The highlight of each course was a weekly or biweekly meeting in BioMOO or DU. There, faculty and students 'met' in a classroom/lecture hall, communicating by text (3). One was waking up in Chicago, one was ending his day in Eastern Europe, one was freezing in Siberia, one was sweltering in the Philippines, but all were together in a room in BioMOO or DU, learning the latest developments in molecular biology or Perl. In the BRI/DBRI courses the full range of MOO-tools was used, including whiteboards, 'tape recorders', 'intercoms', virtual aligners, and so on.

Other examples, with a similar international faculty and student body, and use of MOO format, include the courses 'Principles in Protein Structure', sponsored by Birkbeck College in the United Kingdom, 'Biocomputing', through the Virtual School of Natural Sciences and 'Structure-Based Drug Design' through the Virtual School of Molecular Sciences (see *Table 3*). A current listing of bioinformatics courses may be seen on the Genome Web Site (see *Table 3*).

Advantages of the MOO format for online courses:

- You can attend from office or home, without inconvenience or expense.
- You can whisper a question or comment to the speaker and get a private response, without interrupting the talk, competing for a microphone or attention, or being embarrassed to ask in public.
- With an appropriate client, you can scroll back and catch things you may have missed.

Table 3 Courses, conferences, and virtual laboratories

Courses

IB Homepage	A small international online community of biologists that pursues educational and research collaborations using Internet tools and resources	http:// www.internetbiologists.org/
BRI course homepage (now defunct—see web archive)	BRI is Internet Biologists first course and has had several international editions introducing the use of the Internet to biologists (Web archive 23 January, 2001)	http://web.archive.org/web/20010123085400/ http://www.internetbiologists.org/bri99/ index.html
BRI course content (now defunct—see web archive) DBRI course homepage (now defunct—see web archive)	Description of BRI course content (Web archive 21 April, 2001) DBRI is Internet Biologists advanced course covering topics for Internet-based research (e.g. collaborative systems, sequence-based analyses, molecular visualization, virtual laboratories). There have been several international editions. (Web archive 7 October, 2001)	http://web.archive.org/web/20010421090738/ www.netbio.org/bri99/content.html http://web.archive.org/web/20011007083707/ http://www.internetbiologists.org/dbri99/
Perl for biologists homepage	IB's most recent online course introducing the PERL scripting language from the basics to bioinformatics applications. The first edition in 2002–2003 consisted of lectures/discussions, web resources, and exercises	http://www.ib-perl.org
Lau, C. H., Leifer, Z., Kossida, S., and I. R. Schaffner, Jr. (1997). BRI: a model solution for remaining current on Internet resources and technology. Presented at the Virtual Conference on Computers in University Biology Education	An early article about the BRI course that was the beginning of IB	http://science.uniserve.edu.au/mirror/vCUBE97/ html/chin_hoon_lau.htm

Order out of chaos: organizing your Internet searches by Zev Leifer	A lecture from IB BRI course	`http://emile-21.com/BRI/ BRI97_transcript_19970619.html`
Course: Principles of protein structure	A course presented by the Department of Crystallography, Birkbeck College, University of London.	`http://pps.cryst.bbk.ac.uk/`, now in its 8th year
Biocomputing hypertext coursebook (1996)	Workbook from online course given by the Virtual School of Natural Sciences (VSNS)—BioComputing Division, University of Bielefeld	`http://www.techfak.uni-bielefeld.de/bcd/Curric/welcome.html`
VSNS Homepage	See reference to VSNS above	`http://www.techfak.uni-bielefeld.de/bcd/welcome.html`
Course: Structure-Based Drug Design (defunct site—see web archive)	Certificate/diploma course given by the Virtual School of Molecular Sciences, University of Nottingham (ca 1997) (Web archive 6 December, 2000)	`http://web.archive.org/web/20001206143600/` `http://www.vsms.nottingham.ac.uk/vsms/sbdd/index.html`
The Genome Web. Bioinformatics Course Information	UK Human Genome Mapping Project Resource Centre	`http://www.hgmp.mrc.ac.uk/GenomeWeb/docs-bioinf-course-info.html`

Conferences

Glycoscience Conference	Sponsored by Eurotron, presented by Netconferences. A meeting of glycoscience researchers	`http://www.netconferences.net/eurotron`
Virtual Conference on Computers in University Biology Education (Virtual CUBE97)	An online conference sponsored by CTI Biology, University of Liverpool (1997)	`http://science.uniserve.edu.au/mirror/vCUBE97/home.html`

Laboratories

Leifer, Z. (1997). A microbiology office-laboratory in cyberspace	A paper presented at the vCUBE conference about a virtual microbiology laboratory written in MOO language, previously hosted on BioMOO, now reproduced in DU	`http://science.uniserve.edu.au/mirror/vCUBE97/html/zev_leifer.html`
The Cybertory™ Virtual Molecular Biology Laboratory.	A web-based virtual laboratory written in JavaScript	`http://www.attotron.com/cybertory/`

- With an appropriate client, you can log (record) the lecture and have a permanent record not only of what was formally presented but also of all discussion (including whispered comments and responses).

- With a web/telnet connection, you can go to various websites and have the lecturer discuss the site, take your questions, etc. in real-time.

3.2 Conferences

A number of international conferences have been held online in recent years (see *Table 3*). Netconferences/Eurotron runs meetings on Glycoscience. CTI Biology ran a conference on University Biology Education, using BioMOO for a large group discussion. Others used a chatroom format for online real-time discussion, that is, they lacked the object-oriented tools capabilities of the MOO but had the same whispering/logging/simultaneous-at-a-distance features.

In addition to the above advantages, which apply here, as well, there are some new aspects that apply to conferences.

With an appropriate client (and in some cases web-based software) you can attend multiple sessions simultaneously (try that in real life). That is, when there are multiple sessions in multiple rooms, you can participate in both. The screen indicates when there is new text to read; you can scroll back, if needed, to catch up if you missed anything. You can whisper to the speaker or a colleague 'meet me later, I have some ideas', go to a separate room ('lounge' or 'coffee shop') and develop a research collaboration.

3.3 Laboratories

There are a number of online opportunities for research training. One is the MOO-based 'Virtual Microbiology Laboratory', originally created in BioMOO and reproduced in DU (*Figure 2*); Login to DU (see *Table 2*), then go to the EGL teaching laboratory, @go #20968 (see *Table 3*). There, by typed commands, the student-at-a-distance can manipulate pipettes and petri dishes, doing, virtually, an important experiment, with the instructor-at-a-distance following along and making comments, as necessary. MOO-tools aid in the training process. This is useful as a dry run through a procedure involving pathogens, carcinogens, or radioactivity, where one would rather make costly or dangerous novice mistakes online than in the real lab.

Another is a web-based training experience called the Cybertory, a JavaScript-based virtual molecular biology teaching laboratory (4). A step-by-step protocol in molecular biology is followed, as would be done in the lab. It has no telnet component of its own but it is interactive, realistic and, again, an opportunity to learn a procedure before trying it out with real reagents.

Remarkably, by loading and running the Cybertory program in an independent web browser window, while in constant communication via a telnet client connection on BioMOO, it was possible for several international participants in the DBRI course to be coached by the instructor while collaboratively exploring this teaching tool. Wild type and mutant samples of sickle cell beta globin DNA

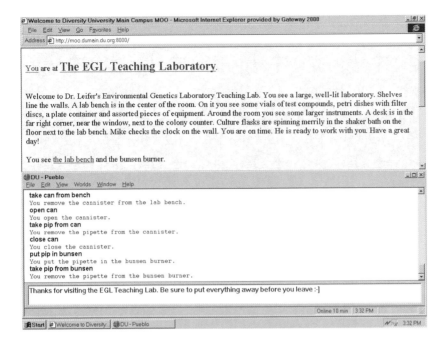

Figure 2 Virtual microbiology laboratory in the DU MOO. A MOO ('Multi-User Dimension, Object Oriented') is a virtual environment for real-time interaction. (See Section 2.2.3.)

were digested by restriction enzymes then fragments separated by gel electrophoresis and visualized by UV, all in virtual space. The same could be done today with a simultaneous web/DU connection.

3.4 Research and project collaborations

Ultimately, the power of these techniques can be best utilized in the area of collaboration between professionals. This can be in the area of biological research or other projects (see *Table 4*).

As examples of the first group, the VIWB sponsors online collaboration. Its website has a number of projects listed and solicits groups to 'find each other' and get together. One such is the Glycoscience Group, which met regularly in BioMOO. Although physically distant, they met and talked and shared ideas that translated to bench or Internet projects. Over the years, pairs of researchers have met in BioMOO to discuss ongoing collaborative efforts.

Perhaps the best example of the second group is IB. The authors of this chapter are all members of IB. Over the last 5–6 years, they have kept in constant contact, via email and in BioMOO/DU meetings, to plan the BRI/DBRI/Perl courses, to plan other IB projects, to work on grant applications, to publish jointly online and to work out internal organizational issues. They have never met but the personal bonds of mutual interest and accomplishment make it as functional and as productive as any time-and-space-bound organization.

Table 4 Research and project collaborations

Leifer, Z. Research Collaboration via MOO	An article in HMS Beagle on the use of the MOO format for online research collaboration (Prior rapid, free registration with BioMedNet is required)	`http://news.bmn.com/hmsbeagle/24/webres/insitu.htm`
Virtual Institute of Experimental (Wet) Biology (VIWB)	A venue to publicize member's projects and a call for collaboration. An IB project	`http://www.md.huji.ac.il/biology`
VIWB Projects	A listing of the members and their projects	`http://bioinfo.weizmann.ac.il:3680/projects.html`
Globewide Network Academy	Distance learning catalogue and related GNA projects. 31645 courses and 4348 programs listed	`http://www.gnacademy.org/`
Overcoming distance and language barriers		
Language Choice Online: Globalization and Identity in Egypt by M. Warschauer and G.R. El Said	Cultural factors affecting online language use and translation	`http://www.ascusc.org/jcmc/vol7/issue4/warschauer.html`
Multilingual Internet: language, culture and communication in instant messaging, e-mail and chat. A call for papers	Announcement in the Journal of Computer-Mediated Communications (at the University of Southern California)	`http://www.ascusc.org/jcmc/cfpmultilingual.html`
Babel Fish, World: translation.	One of the first and best online free language translators by AltaVista	`http://babelfish.altavista.com/translate.dyn`
Systran information and translation technologies Internet interpreters.	Systran software is used by several language translators	`http://www.systransoft.com`
CNET Reviews	The CNET site may be searched for reviews of AltaVista Babel Fish, Transparent Language FreeTranslation, and Translation Experts InterTran	`http://www.cnet.com/`
Open Translation Engine 0.7.9	Open source project for language translation	`http://www.ibiblio.org/dbarberi/ote/stable/`
Google language tools	Translate text or web pages	`http://www.google.com/language_tools?hl=en`
Langenberg.com Language Translation	Another online translator for English, French, German, Danish, Spanish, Finnish, Czech and Italian	`http://translation.langenberg.com/`

WorldLingo.com	Add an online language translator to your browser (subscription) Conduct multilingual chat in 10 languages (demo available)	`http://www.worldlingo.com/products_services/browser_tool.html` `http://www.worldlingo.com/products_services/chat.html`
IBM Research. Translation	Description of the IBM web intermediary (WBI) as a language translation plugin, using the FreeTranslation engine as an example	`http://www.almaden.ibm.com/cs/wbi/doc/examples/translation/`
Language Partners International	Reference Center for foreign language translation	`http://languagepartners.com/reference-center/start-here/index.htm`

Overcoming cultural barriers

Lau, C. H. (1998). In Search of Permanency: A Founder's Perspective on Sustainability of Purely Virtual Team	Article from the founder of IB on virtual teaming	`http://emile-21.com/NetBio/permanency.html`

Collaboration with grid computing

A Grid Computing Primer	Describes the background to grid computing and gives links to other useful resources	`http://www.grids-center.org/grids_primer.asp`
Scientists giddy about the grid by Randy Dotinga. Wired News. 20 January, 2003	Short article touting the possibilities for grid computing	`http://www.wired.com/news/print/0,1294,57265,00.html`
BIRN. Biomedical Informatics Research Network	Introduces the BIRN initiative and describes the motivation and role of grid infrastructure	`http://www.nbirn.net/About/Overview/Objectives.htm`
Scientific Objectives of the Biomedical Informatics Research Network (BIRN)	Describes exemplary uses of BIRN resources in the field of neuroimaging	`http://birn.ncrr.nih.gov/birn/birn_getpage.php?fname=about_birn_bg.html`
About BioGRID at the National University of Singapore (NUS)	Brief introduction to BioGRID	`http://www.bic.nus.edu.sg/biogrid/`
Biology Workbench portal at the San Diego Supercomputer Center	A portal that simplifies access to sophisticated biological research tools (registration required)	`http://workbench.sdsc.edu/`

Protocol 4

Finding a research collaboration

1 Discover collaborators in your area of interest: attend online conferences and courses, offer a short course in your field of expertise, join focused mail lists, follow and participate in newsgroups, use online journals and bibliographic tools to keep current in the literature, visit or join research communities, search online grants databases, and attend online videocast lectures.

2 Contact potential collaborators you have found by e-mail or by striking up a discussion at online conferences or courses.

3 Meet in DU to discuss possible collaborative projects. Over the years, BioMOO has fostered numerous collaborative efforts: DU is available to fulfill that same role. Alternatively, meet in a web-based chat room created for your collaborations.

A similar case could be made for the Globewide Network Academy (GNA). GNA publishes an online catalogue of some 31,000 web-based courses and does much to sponsor online education. Its officers and board of directors may be continents apart but they met in the GNA Forum in BioMOO or Diversity University.

Finally, this chapter itself—the result of online collaboration, mostly by e-mail, archived in Yahoo! Groups (see *Table 2*), the product of a group of diverse professionals, physically distant but as close as professionals and friends can get, linked by the Internet.

Protocol 4 lists some of the most successful approaches to help find a research collaboration on the Internet.

4 Overcoming the barriers to Internet collaboration

Beyond the technical aspects of computing, or ignorance of them, what are the critical barriers to effective Internet collaboration? This section considers three main barriers that may be encountered during a typical Internet collaboration: distance, language, and culture. This dissection of the barriers may seem somewhat simplistic, and other choices may be equally valid. For example, problems with communications, knowledge or understanding, ethics, geography, and administration, to name only a few, may all raise barriers to effective collaboration—the list is almost endless, as are the barriers and respective solutions.

4.1 Overcoming distance barriers

At first glance, the actual distance between Internet collaborators would not appear to matter much; after all the primary modes of Internet communication are almost instantaneous throughout the world. At least, that is what one expects. However, no system is perfect.

The greater the physical distance between 'on-line' collaborators the more likely communication will become asynchronous between them, regardless of the tools used. Every piece of hardware (or software) used to virtually connect any two collaborators has the potential to disconnect or fail at every point in time. For the most part, the Internet, by inherent design, attempts to redirect or re-establish communication lines. Failing that, it may provide appropriate notification of the failure and/or cause. This does not help when completing a critical mission on a tight deadline. Moreover, the collaborators can typically do little or nothing about it except tolerate and understand it. Planning procedures to fall back on when your first line of Internet communications fails is highly recommended.

Collaborators in distant parts of the world may also experience diurnal asynchronicity. In this situation a combination of asynchronous and synchronous communication methods may be necessary. However, diurnal asynchronicity may also be turned to advantage; for example, an e-mail message sent from Washington D.C. in the evening may be received in Singapore at the start of the day and get an immediate reply. Furthermore, when both collaborators are online and net traffic and other factors permit, e-mail may also provide near synchronous communications.

4.2 Overcoming language barriers

Regardless of the number of collaborators and their global diversity, a common language is frequently established early on. However, where there is no one common language amongst the collaborators, translation methods may be used. Unofficially, the global language is English, but this should not be a cause for Anglophones to forget the potential in collaborating with scientists who do not use English as their primary language; good science has no language barrier.

4.2.1 Language translation

To meet the challenge, some collaborators actively pursue learning another language and some are gifted at it. Traditionally, intermediary collaborators who speak two (or more) languages act as translators. Alternatively, there are both web-based and hard drive-based software applications that will attempt language translation on demand (5) (see *Table 4*).

The main drawbacks to machine translation are that the grammatical correctness of the translation may leave a lot to be desired and context is not considered, but the general meaning (i.e. gisting) often can be deduced from the result. This is one area where text-based communications still have an advantage over voice communications.

Cultural factors may also affect the feasibility or quality of machine translation. Diglossia, the differences between spoken and written languages or dialects, frequently occurs in e-mail (especially informal e-mail). Another interesting adaptation to e-mail occurs in Romanized Egyptian Arabic where the numbers

2, 3, and 7 are used to represent Arabic phonemes in the Roman alphabet (see *Table 4*).

Research into online communications in non-English populations has been sparse. The Journal of Computer-Mediated Communications announced an upcoming special issue on this topic at the time this chapter was going to press (*Table 4*), and may be an excellent source for further information.

Alta Vista's Babel Fish, which translates five languages, was one of the first translation tools on the web and still one of the fastest (see *Table 4* for this and other resources below). Babel Fish also has a translation plugin for Microsoft Office applications and browser. Go Translator and Systran Translation Software use the same translation engine as Babel Fish with similar results. FreeTranslation software, while not as accurate as Babel Fish, performed slightly better in tests of longer passages or more complex sentences. Google also provides web-based translation for either text or web pages identified by the URL. If you prefer an open source project there is Open Translation. WorldLingo also provides a subscription translator for your browser or multilingual real-time chat. For a more technical description of web intermediaries and the use of a translation plugin refer to the IBM Research site (see *Table 4*).

Language Partners International provides a reference centre on the types of foreign language translation tools available (see *Table 4*).

4.2.2 Communicating by diagrams

Another method available is to communicate mostly by diagrams, charts, and symbols. This may at first appear to be impossible to accomplish. However, it has its merits, the greatest of which are: (a) a single diagram can convey a vast amount of information on one page, (b) the majority of a diagram is wordless, and (c) studies have shown that the memory retention of pictures is nearly 85–95 per cent accurate.

The main complication with using diagrams is that, by convention and indeed necessity, they will both contain and be accompanied by legend text; first to annotate a symbol, and second to describe the content. However, if combined with the use of a translation tool, this is readily circumvented. In addition, most e-mail applications now support attachments of graphic files as well as text bodies.

4.3 Overcoming cultural barriers

Apart from language, why would a difference in culture be a barrier between collaborators? Mighty tomes have been written on the subject and space does not permit an in-depth study. However, collaboration is by nature a social event where the greatest barriers are likely to result from the ignorance of each other's cultural diversity. The best distillate of advice on the subject is to enjoy singularity and revel in diversity.

In conclusion, learn what you can, adapt as you can, and enjoy it—it's a big world made small.

5 Sustaining voluntary collaborative organizations on the Internet

Most scientific assemblies on the Internet are transient. Geographically distant individuals or research groups may initiate contact for a specific purpose (e.g. to collaborate on a research investigation, writing project, course or workshop), then naturally dissolve their association or become dormant. Other groups choose to maintain their collaborative relationships over many months or years, and may evolve into virtual collaborative organizations with their own agenda and objectives. IB, which had its beginnings as a single collaborative course in 1996, has evolved into an entirely virtual and voluntary, but relatively stable, global collaborative organization. The approaches taken by its founder and members in sustaining this collaboration over the past 5–6 years are provided in *Protocol 5*, as an example for similar collaborative ventures.

Protocol 5

How to evolve and sustain a voluntary collaborative organization in cyberspace—the IB model

1 It is essential to convert the initial collaboration into a simple administrative structure. This is the beginning of a process that will coordinate and channel voluntary resources in a productive direction; such resources may include personal time, intellectual property, collaborative tools, online infrastructure, or endorsements.

2 The administrative structure should distribute the responsibility for development of the collaboration as early as possible. This will enable parallel development of collaborative projects, communication tools, membership, website, sponsorship (if appropriate), etc.

3 Defining and stating the objectives of the collaboration provide a focus for future development. If a consensus can be readily reached amongst its members, the group has a much better chance of maintaining their interest and support. A mission statement should be refined and agreed upon.

4 Membership guidelines help to ensure that growth does not occur at the expense of the original objectives. The objectives will determine the skill sets needed. Creation of new knowledge and innovation will be (or should be) more important to the scientist members than simply membership growth, web site popularity, etc.

5 A procedure for decision-making is essential when critical decisions must be made. Where there is no formal organization or legal entity, establishing an acceptable and effective mechanism to reach a consensus over the Internet can be a challenge. However, this need not be a formalized process so long as there is general agreement.

Protocol 5 continued

6 The nature of any sustainable voluntary organization requires that membership involvement is flexible and that entry/exit strategies are developed for its membership. Members must be able to specify their roles and time commitments, and to change their involvement as time and energy permit.

7 To avoid intellectual isolation and to broaden access to resources and expertise, it is important to continuously develop and nurture partnerships, sponsorships or affiliations with like-thinking organizations.

8 Leadership renewal is both necessary and valuable for several reasons:

(a) The substantial time commitment of voluntary leadership cannot be sustained indefinitely.

(b) Uninterrupted development of projects requires smooth leadership transitions.

(c) Greater stability of a virtual organization can be achieved by having both current and past chairs in its membership.

9 Redundancy in all critical parts of a virtual organization is especially important, since the availability and commitment of voluntary members is subject to many factors often beyond control. Consider redundancy of administrative members, project leaders/coordinators, web server administration, and mailing list administration.

These suggestions for sustaining a voluntary organization on the Internet are based on experience with a successful model, which could be emulated by other organizations. However, in such a complex undertaking there are few if any hard rules that are certain to work, and many new approaches may work just as well, or better.

To prepare you for the challenge we recommend general reading on the principles and practice of running knowledge-based organizations (6, 7). The experience of IB is described in an online paper 'In Search of Permanency' (see *Table 4*). Another example describes leadership strategies in the development of Entovation Network, a purely virtual entity (8).

6 Collaboration in a grid world

While the approaches and tools outlined in this chapter will be suitable to power the routine collaborations of most scientists, additional tools and approaches will be created to effectively collaborate in the developing global grid-computing system (9). Grid computing consists of the hardware and software (i.e. middleware) infrastructure for large-scale integration of high-performance networking applications. The unparalleled computing capabilities of the 'Grid' will not only change the way in which collaboration in biomedical research (and other disciplines) is conducted, but it will fundamentally change the direction of the science. Sharing of computers, storage, and large data sets in a transparent fashion within a high-speed global network will enable large-scale analyses or simulations that can support a new wave of discovery science. Real-time discussions amongst investigators will support these collaborations.

An example grid application is the BIRN, an initiative of the National Center for Research Resources (NCRR) at the National Institutes of Health (NIH), which will allow analysis of data at different levels of aggregation and at diverse sites. The testbed for this project will focus on neuroimaging and will allow multi-scale investigations and pooling of data across multiple sites. Another example is the regional BioGrid centered at the National University of Singapore, which will allow participating institutions to collaborate on bioinformatics projects. The Biology Workbench portal at the San Diego Supercomputer Center provides an example of how portals may simplify access to supercomputing applications within a Grid (see *Table 4* for various Grid web references).

7 Concluding remarks

Explosive growth and development in Internet and World Wide Web communication technologies have enabled the globalization of scientific collaboration, principally in the last decade. This ability to globally exchange large amounts of scientific information rapidly, robustly, and at modest cost has had a major impact in molecular biology, where sequence-based and structure-based investigations have come to largely rely on Internet tools and databases located worldwide.

This chapter has provided an introduction to the communications tools, collaborative applications, and personal experiences that may serve as a starting point to mobilize or invigorate your own research collaborations via the Internet. Other chapters in this book describe the specific research tools and resources you and your collaborators may jointly use. We encourage you to experiment with and evaluate the always changing technologies for communications and resource sharing, explore the links to other articles and resources on collaboration provided here and in the companion web page (see *Table 1*), and pursue new collaborations through advanced online training courses, conferences, and virtual scientific communities and networks. You will have succeeded if you can add new directions, depth or perspectives to your current studies.

Finally, the authors of this chapter are members of IB, an independent collaborative group for the advancement of biological research and academic scholarship through the Internet. Internet Biologists aims to foster an international community of biologists to achieve these objectives through a number of collaborative projects. Our home page provides further information on our membership, collaborative projects, and other helpful resources (see *Table 3*).

References

1. Haynes, C. Holmevik, J. R., (ed.) (1998). *High wired: on the design, use, and theory of educational MOOs*. The University of Michigan Press.

2. Coleman, D. (1997). Groupware: Collaborative Strategies for Corporate LANs and Intranets. Prentice Hall. See outline at: `http://www.collaborate.com/publications/techapp2.html`.

3. Gore-Langton, R. E., Young, E., Fuellen, G., and Glusman, G. (1999). Internet training for biologists in BioMOO. *BioTechniques*, **27**(4), 710–14.

4. Horton, R. M. and Tait, R. C. (1999). A virtual molecular biology teaching laboratory. *BioTechniques*, **27**(2), 298–300.

5. Whittaker, J. (2000). Speaking in tongues. *PC Advisor (UK)*, **63**, 187–92.

6. Sengel, P. M. (1992). The Fifth Discipline. Century Business, London.

7. Patching, A. and Waitley, D. (1996). *The futureproof corporation: practical lessons in leadership and innovation for the 21st century*. Butterworth-Heinemann Asia.

8. Amidon, D. (2000). *Virtual CKO: leading through strategic conversations. Chief learning officers: roles and responsibilities*. American Society for Training and Development—ASTD Action series—Leading Knowledge Management and Learning.

9. Foster, I. (2003). The grid: computing without bounds. *Scientific American*, April, 78–85.

Chapter 8

Laboratory websites: How to disseminate information, make friends, and influence people

Chao Lu

Microarray Facility, The Hospital for Sick Children, 555 University Avenue, Elm Wing Room 10104, Toronto, ON, Canada M5G 1X8.
e-mail: chao.lu@utoronto.ca

James R. Woodgett

Division of Experimental Therapeutics, Ontario Cancer Institute/Princess Margaret Hospital, 610 University Avenue, Rm 0-622, Toronto, ON, Canada M5G 2M9.
http://kinase.uhnres.utoronto.ca

1 Setting up a website: why?

The Internet and World Wide Web were born of the need for scientists to communicate and collaborate. In its early DARPAnet days, the network provided teams of physics and computer scientists with a way to efficiently exchange data and commentary. While the popularization, commercialization, and hype of recent years have exploded the use of this network of networks to virtually anyone with a web-enabled device, its relevance to scientists remains as strong as ever. Scientists and students are less willing to physically visit a library to search for information that can be done more easily and faster at home or in the laboratory. There were about 3 million Internet servers worldwide in 1999 hosting about 8 million pages of web content (1). Of those, about 6% focused on scientific and educational content, with the remainder largely being commercial/corporate. The number has increased three-fold per year since then and shows little sign of slowing.

Traditionally, scientific information has been concentrated within highly processed and moulded publications. Over many years, a complex system of primary journals and review journals emerged that is largely organized by prestige, subject matter, and style. A second publication revolution has been quietly occurring. Until a few years ago, the only means to communicate or expose information to a wide audience was to submit papers to a journal or publishing house. Publication of material of any sort was limited by the channel costs and the processes involved. The advent of web-based publication radically changed this situation. Moreover, the fact that the technology for website hosting and authoring became widespread and of low cost opened up accessibility to essentially anyone with a desire to use this medium.

In a world where you *can* set up a website, the question is: 'Should you?' This depends on whether you have information to share with the rest of the world. Such data might include your research programs, published articles (subject to copyright restrictions), and abstracts. But it may also include information that is peripheral to primary data such as resource collections, hints, methods, contacts, essays and opinion pieces, graphics and tools. A good website can act as a virtual storefront, displaying relevant information about the laboratory, what it does and where it's going. As such, it can be a valuable recruiting tool, providing background information for prospective students, fellows, and employers. Such advertising is often under appreciated, as students tend to be far more web-savvy in using the Internet than established faculty.

One other important application of a lab website is as a public repository for publishing materials that complement your research publications. Journals usually have tight word and page limitations that relegate much of the primary data to supplementary material. Since this had to be requested from the authors, such data used to be difficult to obtain. Lab websites are excellent alternatives for these datasets as exemplified by Dr Pat Brown's site at Stanford University (see Web Resources) which is chock full of raw data derived from microarray experiments. Laboratory websites can also be important learning tools, providing a location to deposit specific course assignments or as a communication tool for listing references and additional material. Posting a calendar on-line can even help members of a lab plan a meeting with their over-traveled supervisor and a web cam might help keep track of who's in on the weekend! The dynamic nature of the Internet allows frequent updating and tracking of continuously changing events.

There are several barriers to setting up a lab website. A few years ago, these were mainly technical (since web-mastering required familiarity with programs with somewhat steep learning curves). Today, most applications involved in word or image processing are web-friendly and can export to the various web-specific formats and most computer operating systems have built-in server software. Most of the remaining barriers are therefore administrative. In the wrong hands, a web server can be dangerous, providing a back-door target for hackers, loss of network security and potential loss of data and confidentiality. As a result of several high profile 'hack attacks' into high security sanctums such as the Pentagon and Microsoft, many systems administrators balk at the idea of a laboratory computer being open to the outside world. Much of the risk can be eliminated by careful configuration but many institutions restrict web servers to a central computer bank located outside of their firewall. Files and pages are added to the server remotely using the GET and PUT commands of File Transfer Protocol (FTP) or similar applications and the researcher does not have to worry about day-to-day maintenance of the computer. The administrator of the web server issues you a name and password and you are able to modify, add, and delete files. The first step in thinking about a lab website is therefore to talk to the computing department to find out what is allowed and what hosting services may be available. You can also set up a laboratory computer as a web server, but you should still talk to the computing administrators to ensure you are in compliance with institutional rules. For most people, using a departmental server is the preferred solution.

Technically, if one has practical knowledge with some of the most popular application programs, such as Microsoft Word, CorelDraw, or Adobe Photoshop, it is not at all difficult to use web authoring and publishing software. Setting up and maintaining a more sophisticated website with elements of dynamic content (such as movies, databases, etc.) requires additional time, programming, and design skills.

A simple series of web pages with text files can be generated from any text editor program. Such a text file is edited in a special format, termed HyperText Mark-up Language (HTML), which allows web browsers like Internet Explorer and Netscape Navigator to 'understand' the structure of the file, paragraphs, positioning of graphics, headings, and hypertext links. In 1995, the depth and complexity of this language was relatively limited. Today, many capabilities have been layered onto the simple framework resulting in a powerful but complicated syntax. Along with this development, HTML editor programs have evolved and now perform most of the calculations under the hood, allowing the author to focus on design and content rather than on mechanics (to a point!).

This chapter introduces the commonly used tools, web design skills, and some basics of web programming languages. It is not intended as a comprehensive review or tutorial on web design, web-related computer programming, or graphic editing. For further information on those topics, the reader is advised to consult specific tutorial and reference books, some of which are not only on-line but free! In addition to the print version, all of the web links included in this chapter have been compiled and can be found at `http://kinase.uhnres.utoronto.ca/weblinx.html`

2 Web authoring and publishing tools

Web page design begins with formatting content using HTML mark-up tags. HTML can be created and edited in any of the text editing programs, such as the simple Notepad or Simple Text applications in the Windows and Mac OS, respectively. With newer version of word processing programs, such as Corel WordPerfect and Microsoft Word 2000/2001, these tags do not need to be typed in manually. Almost any document generated by these programs can be converted into a 'web-ready format' by 'Saving as HTML'. That said, the resultant code is usually a poor substitute when compared with dedicated web authoring packages and this approach is only recommended if you wish to rapidly convert batches of previously written word processing documents. For example, if you covert your CV to HTML, the word processor application will create and embed instructions to allow browsers to view your beautifully crafted document as if you had sent the viewer a paper copy. Highlights, fonts, styles, etc. will be noted and translated into a set of Internet-fluent instructions. It gets a little harder if you have non-text elements embedded within your document. For example, you may have a graphic of your university emblem or a photograph of your institute. These non-text files are handled by reference to the main document. Thus, when the web browser reads an instruction in a document to show a graphic, it is told where the file encoding that graphic is located. It then separately retrieves the file and displays it where the HTML code dictates it should be on the page. Thus, every 'page' on a website has a series of associated files that contain additional information. Converting a document for publishing onto the web may therefore generate several files. When this document is 'up-loaded' to a web server, its associated files must accompany it. In addition, the instructions in the reference document for locating the associated files must be accurate. These instructions are provided in the format of Uniform Resource Locators (URLs) and can be 'absolute' or 'relative'. If absolute, they contain the entire address of the file beginning with the name of the server as well as the directory location of the files. If relative, only the directory locations are included. The browser 'assumes' that the documents are on the same server as the originating document. It's analogous to describing your address to different people. To a person living in the same city you may say you live at 25, Tadpole Street. To a person living in a different country you would add the city, state, and country to the street address. The basic principles of coding a page in HTML and using URLs and hyperlinks are described below.

Netscape Composer, a component of Netscape® Communicator, contains a number of features commonly used in web pages such as HTML generation, image insertion, links, and tables. The finished page and its associated files is uploaded to a web server directly from Composer via its FTP or HTTP capabilities. Amaya is both a web browser and web authoring program for Windows and Unix OS. The program and its source code can be downloaded from the World Wide Web Consortium (W3C) website (*Table 2*). These programs help gather the documents together as well as providing a means to update and exchange documents via the Internet.

In addition to the free software, there are many commercial programs available for web page editing and website management (see *Table 1*). These programs tend to cover the demands of both novices and experts and thus 'grow' along with the author's development. Most allow various views of the page as its being constructed. A 'code view' keeps purists happy and is useful for spotting errors and trimming excess code. However, WYSIWYG (what you see is what you get) displays are much easier and more intuitive to work with. The user introduces the elements of the page and the authoring software generates the source code

Table 1 Web resources

Title	URL	Notes
Online tutorials for web authoring		
The Java Tutorial	`http://java.sun.com/`	Java
Java Programming, Main Index	`http://math.hws.edu/javanotes/index.html`	Java
	`http://msdn.microsoft.com/`	HTML,XML,DHTML, CSS,Design,Security
Adobe Tutorials	`http://www.adobe.com`	Image editing, Design
ahref.com > Guides	`http://www.ahref.com/guides/index.html`	Various
EchoEcho.Com	`http://www.echoecho.com/`	HTML, Java, Flash
HTMLCenter—a web design and development resource	`http://www.htmlcenter.com/`	HTML, Java, CSS,DHTML
Macromedia—	`http://www.macromedia.com/`	Web Design
PageResource.com–by The Web Design Resource	`http://www.pageresource.com/`	HTML, Java, CSS,DHTML, Perl,CGI
W3 Consortium	`http://www.w3.org`	Various
WDVL: A Guide to Creating Web Sites with HTML, CGI, Java, JavaScript, Graphics	`http://www.wdvl.com/Authoring/`	HTML,XML,DHTML,Tools, 3D,Layout
SVG 1.0 Specification	`http://www.w3.org/TR/SVG/`	The w3 standard
Adobe SVG Zone	`http://www.adobe.com/svg/`	Download SVG viewer plug-in
Meta Data		
Resource Description Framework (RDF)	`http://www.w3.org/RDF/`	An XML-based technology for describing information on the web.
The Semantic Web	`http://www.w3.org/2001/sw/`	includes links describing potential applications
Dublin Core Metadata Initiative (DCMI)	`http://dublincore.org/`	Metadata
MedCIRCLE—The Collaboration for Internet Rating, Certification, Labelling, and Evaluation of Health Information	`http://www.medcircle.org/metadata/index.php`	An attempt to rate medical site content.
Movie playback formats		
RealVideo	`http://www.real.com`	
Windows Media	`http://www.microsoft.com/windows/windowsmedia/`	
QuickTime	`www.apple.com/quicktime`	

Table 1 (*Continued*)

Title	URL	Notes
Web authoring reference sites		
The CGI Resource	`http://www.cgi-resources.com/`	CGI
Free-CGI	`http://www.free-cgi.com/freecgi/`	CGI
Cascading Style Sheets	`http://www.w3.org/`	CSS
PNG (Portable Network Graphics)	`http://www.libpng.org/`	PNG, MNG and JNG image formats
Portable Network Graphics	`http://www.w3.org/Graphics/PNG/`	Graphic: PNG
HTML,XHTML	`http://www.w3.org`	HTML,XHTML
Website Design—The HTML Writers Guild	`http://www.hwg.org/`	HTML writer
The Source for Java(TM) Technology	`http://java.sun.com/`	Java
JARS.COM Java Review Service	`http://www.jars.com/`	Java
JavaShareware.com	`http://www.javashareware.com/`	Java
JavaScript.com	`http://www.javascript.com/`	Javascript
ActiveState	`http://www.activestate.com/index.html`	Perl
perl.com	`http://www.perl.com/pub`	Perl
MSDN Home	`http://msdn.microsoft.com/default.asp`	Various
Adeveloper.com	`http://www.Adeveloper.com/`	Various
Adobe Web Products	`http://www.adobe.com/web/main.html`	Various
HTML Help by The Web Design Group	`http://www.htmlhelp.com/`	Various
WDVL: Resource Location	`http://www.wdvl.com/Location/`	Various
A Webmaster Resource Guide: Web-Athoring.com	`http://www.webknowhow.net/`	Various
WebReference.com	`http://www.webreference.com/`	Web design
Web Developer's Virtual Library	`http://www.wdvl.com/`	Web virtual library
XML.com	`http://www.xml.com`	XML
Laboratory web sites		
Laboratory Home Page Directory	`http://kinase.uhnres.utoronto.ca/labdirect.html`	Our directory of lab websites
The Brown Lab, Stanford University	`http://cmgm.stanford.edu/pbrown/`	Microarray data
Woodget lab web links	`http://kinase.uhnres.utoronto.ca/weblinx.html`	Web links from this chapter

Table 2 Web page creation software

Software	Company	website	Note
Authoring and Publishing			
Amaya	W3	http://www.w3.org/Amaya	Free
BBEdit	Bare Bones	http://www.barebones.com	
Dreamweaver	Macromedia	http://www.macromedia.com/	
Freeway	SoftPress	http://www.softpress.com	
FrontPage	Microsoft	http://www.microsoft.com/	Online tutorial available
Fusion	NetObjects	http://www.netobjects.com/	
GoLive	Adobe	http://www.adobe.com/	Online tutorial available
Image Editing			
Fireworks	Macromedia	http://www.macromedia.com/	Trial available
GIMP		http://www.gimp.org/	Free
Graphic Converter	LemkeSoft	http://www.lemkesoft.com	Trial available
ImageReady	Adobe	http://www.adobe.com/	
PaintShopPro	Jasc Software	http://www.jasc.com/	Trial available
PhotoDraw	Microsoft	http://www.microsoft.com/	
PhotoPaint	Corel	http://www.corel.com/	
Photoshop	Adobe	http://www.adobe.com/	Online tutorial available
Animation Software			
Graphic Converter	LemkeSoft	http://www.lemkesoft.com	Trial available
After Effects	Adobe	http://www.adobe.com/	
Flash	Macromedia	http://www.macromedia.com/	Trial available
GIF Animator	ULead Systems	http://www.ulead.com/	Trial available
ImageReady	Adobe	http://www.adobe.com/	
LiveMotion	Adobe	http://www.adobe.com/	Trial available
PaintShopPro	Jasc Software	http://www.jasc.com/	Trial available
PhotoDraw	Microsoft	http://www.microsoft.com/	

automatically, including HTML and even Javascript (see below). Precise placement of elements on a page is a major challenge in HTML authoring. The WYSIWYG programs use a slew of tricks to achieve a result close to the intended design. As a direct consequence, these programs generate significantly more code than conventional authoring programs. Moreover, there are platform dependencies in the way HTML is parsed—particularly within the tricks used by the layout programs.

No matter what application is used, it is always wise to view new pages on a variety of different browsers and operating systems. Differences can be minimized by choosing certain colours that are 'web-safe' and rendered correctly on both PCs and Macintosh computers, or with different versions of the same browser. Font size is a particular issue for Macintosh based viewers due to differences in screen pixel size between PCs and Macs, which result in smaller lettering on the latter platform. An important lesson in web-design is to cater to all audiences. Since a significant fraction of the scientific community uses Macintosh computers (far larger than the general market penetration of ~5%), its best to minimally test your site using a PC under Windows 98, NT, XP, and Mac OS using both Microsoft Internet Explorer and Netscape Navigator. It is also good practice to test at least the last two primary releases of these browsers, since many people still use old versions.

Which web authoring application is right for you? There is no one answer. If you plan to set up only a few pages of personal web information with a public ISP (Internet service provider) or free public web page provider, you may not need any of the commercial software. Online editing and some pre-formatted templates will do the job (one example is the iTools software offered by Apple Computer). As the site becomes more complex (and they always do), it becomes necessary to employ site management tools. These are often integrated into the commercial packages and allow batch testing of links, preparation of files for uploading to the server, synchronization, across-site styling, etc.

Many commonly used features are built into all the popular web authoring and publishing software, such as automatic creation of HTML source code, drag and drop interfaces, and alternation between source code and page views. Since most of these programs have time-limited trial versions, it is worth testing out a couple before buying. Some companies (such as Microsoft and Adobe) provide online tutorial support (see *Table 1*: Web Resources). One important consideration in selecting a commercial application is how well it integrates with other software tools you may own. For example, the Adobe GoLive interface shares many characteristics with PhotoShop and Illustrator whereas Macromedia Dreamweaver shares interface elements with Freehand and Flash. Microsoft FrontPage interleaves well with other Microsoft software (on PCs). As further encouragement, these large publishers offer bundles of complementary applications at significant discount, such as the Adobe Web Collection (GoLive, Photoshop, LiveMotion and Illustrator), Microsoft FrontPage2000/PhotoDraw2000, and Macromedia Web Studio (Dreamweaver/Fireworks/Flash/FreeHand). Each of these solutions provides all of the software necessary for highly professional website construction for around US$1000. However, it is advisable to initially use lower end programs both to save money and for the easier learning curves. One downside to the higher-end packages is the significant cost of upgrades. It's easy to get sucked into a never-ending upgrade cycle where the advances rarely justify the cost, at least for the average scientific web master.

3 Web design

Having access to the best software tools in the world does not help unless you have spent time planning out what you want your website to portray. There are some basic steps to follow as well as some rules of thumb.

3.1 Beg, steal, and borrow

Many sites extol the virtues of various laboratories. Check these out and note the good and bad aspects of their content, layout, and style. Make use of the source code viewing capability of browsers to interrelate code with a particular style or feature. Although there is no absolute way to ensure a well-designed site, you can often decide what works and what doesn't work simply by surfing around. While plagiarism is usually frowned upon, cutting and pasting a few lines of code to get started harms no one and soon evolves into a unique design that, if effective, will be copied by visitors to your site. There are many established websites that invite cutting and pasting of graphics, javascripts, and other elements. Of course, it is both unfair and illegal to directly duplicate certain elements, such as copyrighted or trademarked material, graphics, and overall themes. If in doubt, email the webmaster and ask permission to copy some of their material. Usually the author is honoured and gives consent.

3.2 Keep it simple—at least initially

Organize the information you wish to display in a logical manner. Put contacts and biographical information together, with email links. Group publication lists, information sources, and references. Give special effort to the first 'gateway' page. This is the first impression visitors will have of your site and will likely determine whether they stay and look around or beat a hasty retreat. The 'home' page should define your site, providing information about what the site is about, who is its intended audience, and what this audience may find there. Demonstrate a very early version to a critical associate. It's too easy to become enamoured with your own efforts. Invite criticism. Be prepared to dump pages of carefully crafted code for the sake of clarity. Some ideas just don't pan out.

3.3 Keep it current

Websites accrue content and it is important to refresh that content frequently. Like yesterday's newspaper, web information tends to age quickly. People using the Internet expect to see up-to-date information and quickly move on to less dated content if there is a hint of neglect. Refreshing the information need not be arduous, especially if you have planned for it. Thus, divide up your data by year and shuffle last month's information backwards. Don't delete it, simply overlay newer data and hyperlink it to the older material. Think about the website as a network of interconnected tunnels in an anthill. Some of the pages may be empty (placeholders) awaiting attention. Some may merge into large spaces. No matter what the initial intention, the site will soon become a labyrinth. For this reason it is essential to provide several navigation tools. Don't rely on the visitor clicking on the browser's 'back' button. Provide shortcut links to your home page and to the previous logical page. Some designers use frames to maintain a window with consistent links. Others prefer to keep such links at the bottom or side of each page. Whatever method you choose, be consistent and do not underestimate the likelihood that the visitor will get lost.

3.4 Apply logic—after all you are a scientist!

If the content is useful to you and members of your laboratory, it will likely be useful to others. You have the freedom to include whatever you want and can therefore build into the site the links and tools that you most often use. Links to journals you read, institutions you collaborate with, funding agencies that support your work and sites of colleagues can provide useful starting points. Having these links in one place and close at hand helps increase your efficiency.

3.5 Be approachable

Although the term World Wide Web implies universal access, this is not usually the case. People use different browsers, different version of those browsers, different platforms, different bandwidths, and different languages. It is impossible to cater to everyone but there are some tricks to ensure that most people see your site as you intended. Use 'web safe' colours whenever possible. The web safe palette uses the hexadecimal codes 00, 33, 66, 99, CC and FF values for red, green, and blue (see Web Reference). Do not assume that your visitors will have a large monitor. Screen real estate is expensive and with the trend towards LCD screens and hand held devices, efficient use of space is becoming highly relevant.

3.6 Maintain your site as you would a bicycle

Check for broken or outdated links on your site. These may arise whenever you alter the basic layout, remove an old page, or just make a mistake. Many web authoring applications allow scanning for link integrity and it's a good idea to parse your site through such checks at frequent intervals. External links also evolve and require periodic manual checking.

3.7 Small is beautiful

Minimize the size of your pages and their associated files, especially the home page. Instead of showing every 32 bit photograph, generate a thumbnail size image and hyperlink it to the original image of separate page. Check out your pages by dialing from a modem. If you can't tolerate the wait, neither will your visitors.

3.8 Be consistent

Your site will presumably be hyper-linked to the official university or institute web server. If this is the case, you may wish to retain fluency by adopting the design elements of the parent site. This also keeps you in the good books of the web master. Maintain consistency within your website by using the same background, logos, and buttons throughout. Such uniformity provides a familiar interface for visitors, reassuring them that they are within the same site. Repeated use of design elements also optimizes loading times since web browsers usually cache graphics locally and simply reload from your disk if an image has already been downloaded recently. Some web authoring programs allow the use of template files to aid in overall site design. An added benefit to such approaches is that the overall look of an entire site can be changed by modifying one template file. For example, with Adobe GoLive, the main menu bar, footer note, and copyright can be saved as components, and every page of the site can use them. Any change of the component will trigger an update of every page that utilizes this component, which makes the site management easier and more consistent.

3.9 Be interactive!

In addition to static information content, it is both fun and useful to add interactive web functions to your site such as a calendar listing your course schedule, a simple database for your research group, on-line quizzes, feedback pages, etc. Many of these elements are offered for no charge at various websites and are simple to add and don't require CGI scripting.

3.10 Finally—self promotion

What is the point of creating a website if nobody knows about it? To let people know your website exists you may submit your website URL to popular search engines (some of these charge a fee but others do not). It is best to submit your URL to multiple search engines since no one engine is comprehensive in its coverage. Don't forget to submit your creation to the laboratory web page directory (see *Table 1*: Web Resources). Include 'Meta tags' in the HEAD section of your home page as certain search engines rely on the information in these tags to describe and catalogue your site. Once you have a reasonable prototype of your site, upload it to the web server. It does not have to be perfect and it certainly will not be complete. You'll be modifying the site for the foreseeable future and, by doing so, will be participating in and extending the utility of the World Wide Web.

4 Document publishing

The PDF (portable document format) file format has become the *de facto* standard for exchanging documents via the Internet. In this mature format, people can view and print the document exactly the way it was created, including fonts, images, and text layout regardless of computer platform and operating system. Viewing and printing of PDF files requires Adobe Acrobat Reader, a free download (see *Table 1*: Web Resources).

The easiest way to save a file in PDF format is to 'distill' it using the full commercial version of Adobe Acrobat. Some applications (e.g. current versions of Adobe Illustrator and Corel WordPerfect) offer an option to 'save as PDF' but the degree of control is somewhat reduced. Of note, PDF is a native file format of Macintosh OS X, allowing any application to output as PDF and further entrenching this document creation standard. If you do not have any of these applications but would like to create a few PDF files for your website, Adobe offers a service for generating PDF files online (for a fee). There are additional advantages to the PDF format: files tend to be significantly smaller, the format is 'web friendly' and compatible with IP transport protocols, and there have been no significant instances of viruses being transmitted via PDF files (at least so far). Indeed viral immunity is an excellent reason for converting MS Word documents into PDF files before transmission. The full version of Adobe Acrobat is also useful for scientists who have no interest in web authoring since it allows some control over electronic form filling—a boon for PDF-formatted grant applications. As most journals now archive their papers in PDF format, one can create a virtual library of PDF papers. Adobe Catalogue (part of the full version of Acrobat) aids in organizing PDF collections.

5 Image editing

5.1 Photograph image editing

Image editing has become a common task in most labs due to the flexibility of digitized images for publication, slide preparation, and archiving. While these skills used to be the domain of the graphics/photography department, increasingly, scientists are performing their own image editing. Websites are yet another format for image deposition. There are two basic types of image files: single (still) image files and multiple (animated) image files. Due to the large size of non-compressed images, pictures are typically processed into one of two common compression formats, GIF or JPG. An animated image file is created by linking multiple images together and saving them as a single (usually GIF) file. The GIF format supports 8-bit (256 colour) image files. Images of greater colour depth (e.g. 16- or 24-bit) must be converted to 8-bit before saving as GIF, which often severely reduces the image quality for photographs. For this reason, the JPG format is used for 24-bit images and most digital cameras use this standard to store photographs. JPG allows a choice of compression level. Over zealous compression will decrease the overall image but significant file size reductions can be achieved with no discernible effect on image quality. It is a good idea to use image compression for reasons of efficiency and reduction of drive space (not to mention the viewers who still rely on dial-in modems). In general, GIF images are best used for drawings and simple images, while JPG is best for photographic images. Several image processing applications allow preview of the compressed files so that size can be optimized without loss of overall quality. GIF and its compression technology, LZW, are the proprietary property of Unisys. More recently, a completely open image file format termed PNG (portable network graphics) has been popularized, which can be used for images with colour depths from 1 to 16 bits and is supported by common web browsers.

Specialized software tools are needed for image manipulation, especially in optimization for efficient transmission over the Internet. The clear leader for professional users is Adobe PhotoShop, which offers a comprehensive and powerful series of tools. To address the trend towards content being output to the web rather than expensive printers, PhotoShop is now integrated with another tool termed ImageReady, which provides seamless addition of tools for generating rollover effects, animations, and image splicing. Another strong contender is Corel PhotoPaint, due to its integration with the graphic application, Corel-Draw, thus providing a powerful package for both image editing and graphic drawing needs. Other notable all-in-one editing programs include Microsoft's PhotoDraw (which, typical of Microsoft programs, integrates well with MS Office applications) and Jasc's PaintShopPro. These packages are not cheap and several shareware products are reasonable alternatives, although less integrated. Of these, Graphic Converter (see *Table 2*) is a standout application that allows manipulation and conversion between virtually any image formats. This 'Swiss army knife' program is an absolute must for Mac-based web designers. GIMP (GNU Image Manipulation Program) is a freeware for image editing in Linux/Unix platform (see *Table 2* again).

5.2 Animated images

Addition of dynamic content, in the form of animation, to a web page can increase interest, entertain, or simply be an annoyance. Good design taste is important and restraint is wise. However, in science, animations can be an efficient means of describing a process or pathway. The most common form of animation is via a series of GIF images (animated GIFs) that are composited in a single file and displayed in a defined order. When loaded onto a web page, the browser simply loops at the images.

At a much higher level of sophistication (and price) is Macromedia Flash. This vector-based graphics format describes images as vectors and generates animation by defining changes in those vectors over time. Flash files are very size-efficient since they contain basic image elements along with instructions as to how they are assembled and changed. Flash files are created using either Macromedia's Flash software or Adobe LiveMotion. Playback of Flash files requires installation of a Plug-in or an Active X control. Netscape 4.5 and Internet Explorer 4 or higher include Flash as a component and hence the user usually does not have to download and install the appropriate software. Although Flash/LiveMotion are complicated tools, the results can be spectacular. The interfaces include a timeline, multiple tracks (for background, objects, sound, etc). There are also visualization tricks for planning the track of an object ('onion-skinning'). Indeed, the hardest part of the process is planning what is to be shown, defining the cast of characters (objects), and translating a series of steps into a fluid motion.

A newer vector-based format is Scalable Vector Graphics (SVG). This is an open standard which aims to support many of the capabilities of Flash (see *Table 1*). Of great interest to developers is the fact that SVG is XML-based. As described later in this chapter, using XML can greatly simplify converting information from one format to another, which makes SVG well suited for many data visualization purposes (see also Chapter 11).

5.3 Video

As Internet bandwidth increases, web-streamed video becomes more practical. However, unlike the animation processes noted above, compressed video requires huge amounts of data. Each frame (25 to 30 of which are needed per second for broadcast quality) must be captured, compressed, transmitted, and then decompressed by the viewing computer. As a consequence, Internet video is usually confined to a small window. With digital video (DV) cameras with FireWire/IEEE1394 interfaces getting cheaper, creating and editing video is relatively easy. For content that is intended for web distribution there are a few rules. Reduce movement to increase compression (since compression software compares each frame and is thus more efficient when the differences are small). Keep the video short—no more than a few minutes. Annotate each clip with big text to introduce the subject. Save the movie in at least three different sizes/resolutions and allow your viewer to choose the appropriate version for their connection. Among the compression methods, DivX mpeg appears to generate a smaller file size while maintain good picture quality.

Windows Movie Maker (free with Windows, XP) and Windows Media series offer amateurs easy ways to download, create and edit video. For power users, Ulead's Media Studio and Adobe's Premiere are popular commercial software. There are three major movie playback formats: Real Video, Windows Media, and QuickTime (see *Table 1*). All offer similar quality and are cross-platform. Of note, Microsoft PowerPoint 2001/XP allows slide presentations to be saved as Quick-Time files—an excellent way to 'publish' a set of slides on a website. There are two ways to distribute video on a website. The easiest is to simply place a link to a video file that the viewer downloads to their machine and then views. The downside to this is that all or a very substantial portion of the video file must be downloaded before it will begin to play. In essence, a buffer is accumulated on the local machine. The buffer size (hence, delay prior to playback) depends on the length of the file and the download speed. The second option is to 'stream' the video. This is akin to receiving a television signal. The local computer displays the video but does not save any of the streamed data to disk. One advantage of streaming is that the viewer can 'tune in' to any part of the broadcast and can skip forward or back without having to download the entire file. However, streaming capability usually requires a dedicated streaming server and expensive software (except QuickTime) and is therefore not usually practical for casual websites.

6 Computer languages commonly used for the World Wide Web

The World Wide Web is built using a variety of computer languages. These include 'markup languages' (such as HTML and XML) for organizing and/or presenting data, as well as programming languages (Java, Javascript, Perl, etc,) for providing dynamic behaviour. HTML is the predominant medium for presenting web pages. It can be created either as 'static' pages, as when you save a Word document to HTML, or as 'dynamic' pages created on demand by programs. Static pages are a platform for one-way information flow, while dynamic transactions provide two directional communication, such as for database queries and submissions. We begin with a discussion of HTML because it is the final format for most web pages, whether static or dynamic. Though excellent tools are now available for creating static HTML, which relieve authors of much (though perhaps not all) of the burden of dealing with the details of HTML, it is nonetheless valuable to learn the basics of this language which so pervasively influences the Internet. A brief study of HTML also serves as a gentle introduction to the more general Extensible Markup Language (XML), which is becoming crucially important for interchange of structured data in general, including genetic and other biological data. Finally, we briefly introduce some popular web-related programming languages and describe how they are commonly used in a web-based client-server architecture, since any claim to computer or Internet literacy requires at least a passing familiarity with programming.

6.1 HyperText Markup Language

HTML is a non-proprietary format based on SGML (the Standard Generalized Markup Language) and is standardized by the World Wide Web Consortium (W3C) for the World Wide Web. Converting documents into HTML format is relatively straightforward with a plethora of freeware and shareware tools. A number of excellent commercial applications make page creation relatively painless and even provide guides for implementing complex features that are sure to impress your friends. Indeed, there is a strong tendency to migrate from simple, information-rich pages to visually dynamic but slow-loading (over a modem) environments. Don't forget; though the medium is partly the message, your content is what will determine the usefulness of the website.

HTML was designed for storing and exchanging information between interconnected computers and to support embedding of hypertext links and non-text media (such as images) in documents. The language is quite mature, with changes being incremental and largely dealing with inclusion of new formats and resolving previous ambiguities. Browser applications are reasonably fluent in interpreting HTML layout, although the results are sometimes unexpected.

HTML is embedded in all web pages and can be viewed when you choose 'view:source' (or just click the right mouse button, if you have one, over a link to bring up a context-sensitive menu with this option). The source code for a web page comprises a listing of text fragments marked with bracketed 'tags'. HTML comprises a lexicon of tags describing the purpose of data elements and/or how they should be presented. Tags are generally used in pairs, such as '<html>' and '</html>'. The first is the opening tag that defines the beginning of an element;

the element continues until an appropriate end tag (the one with the forward slash) is encountered.

The essential components of your web page will look like this:

```
<html>
    <head>
    <title>My web page title</title>
    </head>
    <body>
                    Web page content.
    </body>
</html>
```

The bulk of the content that is intended for the user to view is inserted between the <BODY> and </BODY> tags. This content can be made up of combinations of text paragraphs, tables, images, animations, hyperlinks, movies, etc.

6.1.1 Hyperlinks

Hyperlinks are the most important feature of HTML. These links tie together disperse pieces of information, essentially putting any data within one click of a mouse, regardless of the location of the computer hard drive on which that data sits. The hyperlink function is defined with the <a> tag (for 'anchor') which includes a Uniform Resource Locator (URL), the web address of the linked information. Let's say you would like to add a link to *Nature* on your page.

```
<a href="http://www.nature.com/">Nature</a>.
```

In English, this tells the browser to load (hyper-reference) a page at the particular URL. The descriptor is simply the text (usually underlined and coloured blue) upon which the user must click to activate the link. In addition to text, an image can also be embedded as a hyperlink by use of an <img> tag. Use of 'border=0' will eliminate the blue edge of the image. You can use a logo image as a link to the journal.

```
<a href=''http://www.cell.com/'' border=0>
    <img src=''cell\_logo.gif''></a>
```

6.1.2 Tables

The Table tag was originally devised to allow browsers to display data in tabulated form. Web designers quickly stretched its formatting capabilities such that most complex sites control their layout using various nested table tags. Table formats start and end with <table> and </table> tags, respectively, and column and row are defined with <td> and <tr> tags.

```
<table>
    <tr>
    <td>first column of first row</td>
    <td>second column of first row</td>
    </tr>
    <tr>
    <td>first column of second row</td>
    <td>second column of second  row</td>
    </tr>
</table>
```

The sizes of columns and rows can be either relatively or absolutely defined by height or width. Cells can be merged into a bigger cell horizontally with 'colspan' or vertically with 'rowspan' functions.

6.1.3 MetaData

The wealth of data available on the Internet brings to light the importance of 'meta data', or 'data about data'. This is descriptive information that can be used by search engines to identify, index, and catalogue the contents of websites. The HTML 'meta' tag provides one way to include a limited amount of simple descriptive information in a web page, to help search engines index the page effectively. Nested within the HEAD element, this information is invisible to most users but embeds information about the content and coverage of a particular page. Meta tag keywords and descriptions are supposed to present a meaningful summary or list of keywords for the given website or web page. The meta tag lets you define properties such as 'author', 'copyright', etc., and give them values. A set of such properties is called a profile; use of a standard profile can make it easier to provide consistent indexing. One such profile is provided by the Dublin Core Metadata Initiative (http://dublincore.org).

Many search engines store all the words of page content in their database. Meta tags let you add words, such as similar keywords or synonyms that might be useful for searching, even if they do not themselves appear in the document. Unfortunately, inappropriate metatags are common due to attempts by unscrupulous web designers to increase the number of times their sites are found by search engines, even when the keyword is irrelevant to their site ('Spamdexing'). This is one reason that it might be preferable to separate meta information from the content itself, allowing, for example, searches based on third party reviews or classifications of websites. Metadata itself can be quite rich; for example, it could contain information about the metadata source (e.g. the author of a review, or the organization providing a rating) as well as about the data itself. To be useful, it must be machine-readable. To this end, the Resource Description Framework (RDF) language (yet another application of XML, see below) is being developed to implement the 'Semantic Web'. Interested readers are referred to the sites listed in *Table 1*.

6.1.4 Cascading Style Sheets (CSS)

CSS1 comprises a simple set of definitions that allow authors to attach style (e.g. fonts, colours, and spacing) to HTML documents, much the same way that style sheets are used in word processing applications to structure the appearance of text in an orderly manner. CSS2 is a superset of CSS1 and supports media-specific style sheets for presentation to visual browsers, aural devices, printers, and handheld devices. CSS can be written and edited with any text editor software, and can be used in both HTML and XML structural document formats.

CSS makes it possible to control many presentational aspects of a page by specifying details about an element's style. For example, removal of the underline on a particular link can be achieved with an 'in-line style' (i.e. a style applied directly within a tag):

```
<A ref="http://www.nature.com/"  style="
      text-decoration:none">Nature</a>
```

However, the real power of styles becomes evident when they are applied in a 'style sheet', rather than in-line. Style sheets allow the separation of the style and layout of HTML file from the information content. With style sheets, one can control the layout of an HTML page from a single location, which can be at the top of an HTML document in the 'Head' section (embedded), or in a separate style sheet file (external). The latter arrangement allows the overall appearance of a website to be radically altered with relatively few changes.

For example, if one does not want hypertext links to be underlined, an embedded style sheet can remove all underlines from the links in a page with the following settings:

```
<style type="text/css">
<!--
A:link {text-decoration:none;}
-->
</style>
```

External style sheets are an efficient means to configure the style attributes of an entire site using a single file. Style information is linked by a LINK element in the HTML <HEAD> section of each page on the site.

```
<HEAD>
    <link rel="stylesheet" href="text_style.css"
        type="text/css">
</HEAD>
```

In addition to controlling styles based on HTML tags, various parts of a page can be grouped for styling purposes by assigning them an identifier, or putting them in a 'class'. As an example, the CLASS attribute can be added to an element as follows:

```
<A ref="http://www.nature.com/"
 class="navlink">Nature</a>
```

In this way, those links in the 'navlink' class can be addressed in the style sheet separately from other links.

6.1.5 Dynamic HTML

DHTML is not by itself a language, but a collection of technologies including HTML, CSS, positionable elements, and scripting languages, such as JavaScript or VBScript. With DHTML, one can create animation to move elements or exchange layers. Due to some differences in browsers' Document Object Models (DOMs), DHTML based scripts may not work consistently in all browsers, or in every version of a particular browser. While such differences are common with DHTML, they also occur in other web page elements due to different interpretations of standards by browser software coders. Browser development is also dynamic and each new version handles HTML with subtle differences.

6.1.6 A starter laboratory web page

After all that has been said about web programming languages, it may not be a good idea to use all the functions in a particular web page. Use only what is necessary for you. Below is a sampling of HTML code that can be modified and upgraded to generate a simple web page.

```
<html>
<head>
<TITLE>Molecular Biology in My Town: Home page</TITLE>
<META NAME ="keywords" CONTENT ="genomics, dna, proteinomics, genetics,
    genome">"
<META NAME ="description" CONTENT ="Molecular Biology Lab">
<style type="text/css"><!--
.navlink { color: blue; font-weight: bold; font-size: 10pt; font-family:
arial, helvetica, "ms sans serif", sans-serif; text-decoration: none }
.noline { color: blue; text-decoration: none }-->
</style>
</head>
<body>

  <table width=450 height=440>

        <tr>    <td height=25 nowrap width="151"><a href="index.html"
                   class="navlink">Home Page</a></td>
                <td height=25 nowrap width="150"><a href="labmember.html"
                   class="navlink">Lab Members</a></td>
                <td height=25 nowrap width="150"><a href="research.html"
                   class="navlink">Research Interests</a></td>
        </tr>

        <tr>    <td height=25 width="151"> </td>
                <td height=25 width="150"> </td>
                <td height=25 width="150"> </td>
        </tr>

        <tr>
                <td colspan=3 height=25 align="center">
                   <h1>Welcome to our web page</h1></td>
        </tr>
```

```
        <tr>      <td height=25 colspan="2">Our Recent Publications:</td>
                  <td rowspan=5 width="150"><img src="myphoto.gif"
                      name="TheBoss"></td>
        </tr>

        <tr>      <td colspan="2"><a href="my\_latest\_nature\_paper.pdf"
                      class="noline">Molecular biology in Nature</a></td>
        </tr>

        <tr>      <td colspan="2"><a href="BestData.pdf"
                      class="noline">Best Data </a></td>
        </tr>

        <tr>      <td width="151" height="25"> </td>
                      <td width="150" height="25"></td>
        </tr>

        <tr>      <td colspan="2">Journals we like to read</td>
        </tr>

        <tr>      <td width="151"><a href="http://www.nature.com/"
                      class="navlink">Nature</a></td>
                  <td width="150"><a href="http://www.cell.com/"
                      class="navlink">Cell</a></td>
                  <td width="150"></td>
        </tr>

        <tr>      <td width="151"><a href="http://www.sciencemag.org/"
                      class="navlink">Science</a></td>
                  <td width="150"><a href="http://www.emboj.org/"
                      class = "navlink" > EMBO Journal</a></td>
                  <td width="150"></td>
        </tr>

        <tr>      <td width="151" height="25"> </td>
                  <td width="150" height="25"></td>
                  <td width="150" height="25"></td>
        </tr>

        <tr>      <td colspan="3" align="center">Our Research Interests</td>

        </tr>

        <tr>      <td width="151" height="20">DNA methylation;</td>
                  <td width="150" height="20">DNA microarray</td>
                  <td width="150" height="20">Molecular Biology
                      and the Internet</td>
        </tr>

        <tr>      <td colspan="3">Name: John Smith, Ph.D.<br>
                              Research Area: Molecular Biology
                              Internet Applications <br>
                              Address:<ul>
                              <li>Room 5-555
                              <li>555 University Avenue, Metropolis,
                              Neverneverland.
                              <li>Phone: (555) 555-4444</ul></td>
        </tr>
    </table>
  </body>
</html>
```

6.2 Extensible Markup Language (XML)

Standardized in 1998 by W3C, XML is a simplified subset of SGML. Like HTML, XML is a text format and makes use of tags (words bracketed by '<' and '>')

to transmit specifically structured data. Many types of commonly encountered data, from spreadsheet or database tables to descriptions of phylogenetic relationships, is 'structured' in some way. XML is not HTML, nor a replacement of HTML. XML provides a markup mechanism to impose constraints on the layout and logical structure of documents. The power of XML is that it can be extended in straightforward yet flexible ways. In essence, a web designer could create a completely new set of tags and have XML-capable browsers render the code as the designer intended. This is achieved by the use of document type definitions or DTDs. Each new 'tag' has an associated DTD that defines the function and syntax of the tag. This structure allows any type of data to be intelligently identified and interpreted, a particularly important development for science-based websites.

6.2.1 XHTML

Released by W3C, XHTML repackages the HTML into an XML compatible form. The differences from HTML seem minor (quotes are required around attribute values, tag names are case-sensitive, all element start tags must have an accompanying end tag or use special 'empty tag' notation, etc.), but being XML compliant makes the data usable in a wide variety of XML tools, including XSLT style sheets (see Chapter 11) and the XML parsers available for most programming languages. XHTML will be important for the next Internet growth spurt, where it can help meet the demands of 'non-classical' Internet access devices, such as cell phones, small wireless communication devices, and TVs, as well as desktop computers.

Protocol 1

Changing HTML to XHTML

One of the goals of the version 1.0 of XHTML is to form a bridge between the 'old-fashioned' HTML we know and love to the XML-based web of the future. By following certain guidelines, it is possible to write (or re-write) web pages in XHTML that can be viewed in older browsers that were only designed to handle HTML. This protocol shows the major steps for converting a web page from HTML to XHTML by hand. Writing web pages in XHTML can be an exercise for learning XML. For more specific details, refer to the sources in *Table 2*.

1 Make sure all tag names are lower case. In HTML, tag names could be written in upper case, lower case, or mixed. XHTML, however, is case-sensitive.

2 Don't use overlapping tags. In old fashioned HTML, you could write tags like this: <B>bold <I>and italic</B></I>. Note that the bold section (<B>) overlaps the italic (<I>) section. In other words, the bold section starts first, then the italic section, then the bold section ends, then the italic section ends. In XHTML (and XML in general), this kind of overlap is forbidden. Sections must be 'nested' inside of other sections. This example would be re-written as bold <i>and italic</i>.

Protocol 1 continued

3 Be sure all start tags have corresponding end tags. HTML allows list items, for example, to be marked by the <li> tag; in XHTML you must explicitly mark the end of a list element with an end tag, like this: <li>an item on a list</li>. Empty elements, like the Horizontal Rule tag (<HR>), are those that contain no text or other elements inside them. In XHTML you must mark the end anyway, (<hr></hr>), or use a special abbreviation for empty tags that uses a slash at the end of the tag to show that it is an empty element (<hr/>). In XHTML, it is recommended to put a blank space before the slash, so it doesn't confuse old HTML browsers. The space is not generally required for XML. In XML, this special empty tag notation can be used for any tag that happens to be empty; to make XHTML that can still be understood by old HTML browsers, it should be used only for tags that are empty by definition. For example, to mark a 'paragraph' that has nothing in it, you should use '<p></p>' rather than '<p/>'.

4 Put quotation marks around all attribute values. Quotes are optional for attributes in HTML, but not in XHTML. For example, in the tag <body bgcolor="red">, the quote marks around the word 'red' are required.

5 Use external files for scripts and stylesheets. It is possible to include them directly, but they are handled differently than in HTML, and linking to external files will help avoid surprises.

6 Validate your XML on the web at `http://validator.w3.org/`.

Detailed guidelines for creating XHTML that is backwardly compatible with HTML browsers are given at `http://www.w3.org/TR/xhtml1`.

(Protocol by R. Horton)

6.3 Java

Java, developed by Sun Microsystems, is an object-oriented programming language that resembles a simplified form of the C++ language. Java was designed to be platform agnostic; that is, to work on any computer or operating system. Unlike most of the common languages that preceeded it, Java is normally compiled into a form that runs on a Java Virtual Machine (JVM), rather than on a particular type of central processing unit (CPU). In theory, if the JVM is available for a particular computing platform (CPU and operating system), then Java programs should run on that platform. Different operating system and software developers have only partially embraced this vision such that 'write once, run anywhere' is still more of a promise than reality. The central concept of object-oriented programming is the object, a module containing data together with the subroutines that act on that data. Objects provide very powerful approaches to organizing programs, and to making parts of programs reusable. Java programmers display graphics, access the network, and interface with users all by using or adapting predefined objects that form part of the 'Run Time Environment' on the JVM. In addition to full-blown applications (standalone program that can be used for general purposes), Java can be used to write 'applets', small programs usually embedded within other applications (such as web pages) to provide a particular functionality. The viewer has to use a Java-compatible browser for such applets to run. While most browsers are Java-capable, some users turn off the feature as it has been linked to security issues. The Java language is currently in version 2, which runs on version 1.2 and above of the Java Runtime Environment (JRE). The JRE is in version 1.4.2 as of this writing. To learn more about Java and Java programming, the reader is referred to rich web resources and reference and tutorial books. The Sun Microsystems Java website contains information for both beginners and professional Java developers.

It is possible to take advantage of Java functionality on web pages without detailed knowledge of the structure, classes, and programming nuances of Java. Many web authoring tools have built-in Java programming support to automatically write the necessary code when certain Java functions are selected. Many Java applets are available for free on the web for effects that include animation, rollover buttons, message boards, pop-up menus, animated banners, chart boards, etc. Installation of these Java applets is usually not difficult, and many applets have accompanying instructions.

6.4 Javascript

Javascript is, confusingly, not Java but instead is a scripting language that was developed expressly for use in web pages. Javascript is simpler than Java, and many consider it easier to learn.

The following script 'pre-loads' images for a rollover effect.

```
<script language="JAVASCRIPT"> <!-
if (document.images) {
image1 = new image
image2 = new image
image3 = new image

image1.src = ``first photo.jpg"
image2.src = ``second photo.jpg"
image3.src = ``third photo.jpg"
}
//-- >
</script>
```

In this example, the name of image is 'myphoto' and placing the mouse cursor over the image changes the source of the picture displayed:

```
<a ref=# onmouseover="document.myphoto.src=image2.src"
onmouseout="document.myphoto.src=image1.src"
```

```
onclick="document.myphoto.src=image3.src">
<img src="first photo.jpg" name="myphoto"></a>
```

A rollover effect can also target a 'remote' image, that is, the image can be changed when the mouse cursor touches some other part of the page.

```
<img src="first photo.jpg" name="myphoto">

<a ref=# onmouseover="document.myphoto.src=image2.src"
onmouseout="document.myphoto.src=image1.src"
onclick="document.myphoto.src=image3.src"y>
Change Remote Image from This Text </a>
```

6.5 Server-side programs and Common Gateway Interface

The World Wide Web is based on a 'client-server architecture', in which users (usually on a personal computer) make requests over the network for a program running on another computer to do something. In this scenerio, the user's computer (or, more precisely, the web browser program running on the user's computer) is a 'client', and the remote computer (or a program on the remote computer) is the 'server'. This arrangement is very flexible; a single client can make requests of many servers, and each server can reply to many clients, in a truly web-like fashion. A single physical computer can run both client programs and server programs, which may be communicating with each other or with other programs on the network. Not all clients are web browsers; other programs such as website indexers or intelligent agents can take on the client role.

Probably the most familiar server programs are web servers (such as the open source Apache server, or Microsoft's Internet Information Server) that deliver HTML files on request, but there are many other services in common use. Unlike static pages with fixed contents, pages created on demand by programs can output 'dynamic' information as requested by web users.

The Common Gateway Interface (CGI) is a standard way to permit interactivity between a remote client and a program running on a server computer, using a web server as an intermediary. A huge variety of programs can be used through CGI, including shell scripts, Perl scripts, and compiled Fortran, Pascal, or C or C++ programs. The CGI protocol specifies the interface by which the web server program passes information requests to another program (the 'CGI' program) and returns the resulting output (usually in the form of an HTML document).

Note that not all server side programs are CGIs. Many other approaches exist for extending the functionality of a web server, including Active Server Pages (ASP), Java Server Pages, PHP, mod-perl, etc. Rather than having the web server communicate with other programs, these technologies typically allow the web server process itself to handle requests dynamically. This approach generally provides a performance boost over CGI.

Probably the most popular application for server-side programs is to handle requests for database searches (PubMed, BLAST, Google, etc), bulletin board news postings, web-based e-mail, etc. The search request is typically submitted from an HTML form and processed by a server-side program which mediates between the client and the database. After selecting the appropriate information from a database (which is usually not directly accessible to the client), the server side program typically formats it in HTML and sends it back to the client's browser for display. Such interactive communication allows effective and efficient delivery of information.

Besides form handling, server side programs can be used for purposes such as web page hit counting and traffic monitoring, etc. Setting up a program that runs on the server computer requires the permission of the system administrator. An easy work-around that can bring many of the benefits of server-side functionality without having to run special programs on your own computer is to take advantage of services provided by Internet companies, such as website statistics, web page counter, or bulletin boards. Some of these are free while others require a fee.

Among programming languages, Perl has become the most popular choice for server-side programming. Perl is an interpreted language developed and optimized for scanning text files, extracting information from them, and printing reports based on that information. It is relatively easy to understand and to use, but is rich in functionality, combining some of the features from older programming languages, such as *C*, *sed*, *awk*, and *sh*. Perl can be used for database connectivity, network programming, web programming, handling e-mail, news, etc. With the Perl/Tk extension, it can also be used with a graphical user interface. Perl database interfaces are widely used. Many websites provide detailed instructions and tutorials on Perl and Perl scripting although these aren't for the faint of heart (see web references). See Chapter 11 for a biologically oriented introduction.

7 Concluding remarks

Publishing a laboratory website has many benefits. It increases your lab's visibility, provides customized tools close to hand, helps other scientists and students to find information, and encourages interaction and collaboration. A website is a form of publication. As such, it demands care, attention to detail, accuracy, and insight. Scientific websites also cater to much wider audiences than scientific papers. Casual browsers or people re-directed from search engines will often be less narrowly focused than the readers of your scientific papers. You may choose not to cater to these visitors but that is a lost educational opportunity. Some of those visitors may be researching a disease they've been diagnosed with or may be kids researching their homework (your site might inspire a career choice!). Other visitors may be journalists looking for background on a new research finding. Yet others may be scientists in different specialties hoping to find an expert or potential collaborator with whom to exchange ideas. By opening your laboratory to the world via the WWW, you assume responsibilities that many scientists do not usually concern themselves with. It's your choice to whom your site caters but, like a popular restaurant, it's a good idea to serve a wide variety of dishes. Open communication is the axle grease of science and the web makes it easier than ever to disseminate information and knowledge.

Reference

1. Lawrence, S. and Giles, C. L. (1999) Accessibility of information on the web. *Nature*, **400**, 107–9.

Chapter 9

Introduction to macromolecular visualization

Eric Martz

University of Massachusetts, Amherst, MA, USA.
www.umass.edu/molvis/martz

Timothy Driscoll

molvisions.com, Waltham, MA, USA.
www.molvisions.com/

1 Introduction

The ability to view a macromolecule in three dimensions is a great asset to science students and researchers alike. A decade ago, the primary method for examining 3D molecular structures was stereoscopic viewing, which often required training or special glasses. Now a number of excellent computer tools, many of them free, are available for rendering molecular structures in 3D, interacting with them, and customizing their appearance. Some are stand-alone programs like WebLab ViewerPro and DeepView Swiss-Pdb Viewer. Others are browser plug-ins like Chime. Still others, like Protein Explorer, a browser-based package using the Chime plug-in, can work entirely over the Web. This Chapter explains which of these programs to use for your particular objectives. However, the scope of this Chapter is limited—it does not cover some other freeware visualization packages with very strong features, including Cn3D, DINO, MAGE, PyMOL, and VMD (find links in the World Index, see the WebLinks box below).

Protein Explorer (PE) is the major focus of this chapter. For basic visualization of 3D protein and nucleic acid structure, PE is the easiest-to-use software, and it is free (1, 8). It is much easier to use than the deservedly popular program RasMol, because you don't have to learn a teletype-style command language. It is also considerably more powerful. The first view of a protein that PE shows you is maximally informative, and is explained in detail (*Figure 1*). As you zoom in, hide portions of the molecule, and change the rendering and colour scheme of the visible portions, a detailed explanation is automatically shown for each

Table 1 WebLinks

Major on-line resources mentioned in the text are gathered here.

Title	URL	Notes
Protein Explorer		
Protein Explorer	proteinexplorer.org	Free, user-friendly software for qualitative visualization of macro-molecular 3D structure with inte-grated knowledge base.
Netscape 4.8	www.netscape.com	Version 4.8 recommended for Pro-tein Explorer. Look for "Prod-uct Archive", then "Browser Cen-tral".
Chime	www.mdlchime.com	Free plugin required by Protein Explorer. See related resources in *World Index**.
Additional Visualization Software **(For many more alternatives, consult the** *World Index**)		
Accelrys Viewer	www.accelrys.com	Formerly WebLab Viewer. Good for publication-quality graphics.
DeepView	www.expasy.ch/spdbv/mainpage.html	Also called SwissPDB-Viewer. Free with powerful modeling capabilities. See DeepView section of *World Index**.
RasMol 2.6	www.umass.edu/microbio/rasmol	This site provides extensive help for beginners. Less powerful and user-friendly than Protein Explorer.
RasMol 2.7	www.rasmol.org	Enhanced version of RasMol.
Operating System Compatibility		
Using Protein Explorer in Linux	molvis.sdsc.edu/protexpl/platform.htm	Enable Protein Explorer to work in linux and other operating sys-tems.
Win4Lin	www.trelos.com	Inexpensive support for Microsoft Windows®within linux.
VMWare	www.vmware.com	Highest-quality support for Microsoft Windows®within linux.
Citrix Metaframe	www.citrix.com	Provides Microsoft Win-dows®desktop (from remote Windows server) within SGI/Irix. Sun/Solaris, linux, and other unix systems.
Molecules (Sources of atomic coordinate files. See also the *World Index**)		
Atlas of Macro-molecules	proteinexplorer.org	Illustrated atlas with links that display molecules in Protein Explorer. Click link to Atlas on Protein Explorer's FrontDoor.
PDBLite	pdblite.org	User-friendly simple key-term searching of the Protein Data Bank for beginners.

Table 1 continued

Title	URL	Notes
Protein Data Bank	www.pdb.org	Primary international archive of all published macromolecular 3D structures with powerful search interface.
OCA	http://bioinfo.weizmann.ac.il:8500/oca-bin/ocamain	Powerful alternative search interface for the Protein Data Bank.
Educational & Scientific Resources and Tutorials		
*World Index of Molecular Visualization Resources	molvisindex.org	Visitor-maintained indices to subjects and authors of diverse on-line molecular visualization resources, including hundreds of tutorials, some in Spanish, French, German, Portuguese, etc.
Chime Resources	www.umass.edu/microbio/chime	Tutorials on DNA, hemoglobin, antibody, MHC, etc. and help for authoring Chime-based tutorials.
EMail Discussions for Molecular Visualization	www.umass.edu/microbio/rasmol/raslist.htm	See others in the *World Index**.
Biochemistry in 3D	www.worthpublishers.com/lehninger3d/	Interactive tutorials covering protein architecture, nucleotides, G proteins, bacteriorhodopsin, oxygen-binding proteins, lac repressor, restriction endonucleases, hammerhead ribozyme, and MHC. Detailed and designed for effective pedagogy.
ConSurf	consurf.tau.ac.il	Automated display of evolutionary conservation for each residue in a protein.
History of molecular visualization	www.umass.edu/microbio/rasmol/history.htm	Early computer models, Kinemages, RasMol, Chime, Byron's Bender, Fred's Folly, physical models by rapid prototype engineering, and molecular sculpture.
Standard color schemes: DRuMS	www.umass.edu/molvis/drums	Standardized, intuitive color schemes recommended for macromolecular tutorials. Freely downloadable.
Structural Alignments	cl.sdsc.edu/ce.html	Help is available under "alignment" in Protein Explorer's Help/Index/Glossary.

If the weblink you are looking for is not listed in the table above, the best place to look for it is the World Index (listed above*).

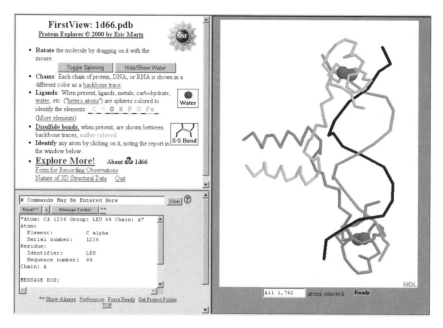

Figure 1 FirstView (grayscale rendering of colour screenshot): the first view of a molecule shown by PE is maximally informative. This example shows the DNA-binding domain of yeast gal4 bound to DNA (PDB ID code 1d66, used in PE's Quick-Start and Tutorial). One can discern the number of chains (the missing colours help), ligands (here, four zinc ions as dark balls), disulfide bonds (none are present here), and water (which has been hidden here). Clicking on an atom reports its identity in the message box (lower left). Although a command entry slot is always available, no commands need to be learned to explore with the QuickViews menus and buttons (see *Figures 3 and 4*).

operation, and a troubleshooting guide is offered. Colour schemes are always accompanied by a key in colour-matched text. A clickable sequence display relates sequence to 3D structure (see *Figure 2*).

Much effort in PE's design has been targeted to novices, occasional users, and non-specialists. Nevertheless, crystallographers and protein structure scientists have praised the image clarity and ease of use of PE's advanced features (*Figure 3*), and compared them favourably to commercial visualization software. Notable are the single-click availability of noncovalent bonding interactions ('Contact Surfaces, see *Figure 4*), the Noncovalent Bond Finder, and the ease with which a 3D protein structure can be coloured to show regions of mutation or conservation from a multiple sequence alignment (1, 9; *Figure 5*).

Quick Start. PE is designed to be self-explanatory. If you want to jump right in, and if you already have Chime installed in your web browser, go to www.proteinexplorer.org and press on the big Quick-Start link at the top. If you don't have Chime, download the installers (see WebLinks below). An on-line guided tour is provided at the big 1-Hour Tour link on the FrontDoor of PE.

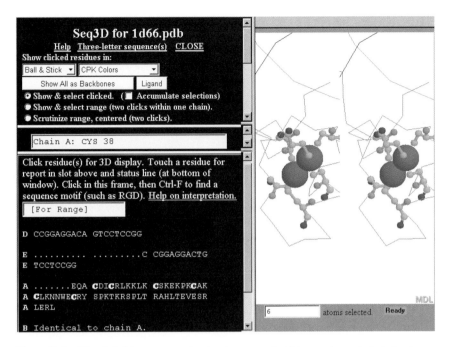

Figure 2 Seq3D: clickable sequence to structure mapping (grayscale rendering of colour screenshot; divergent stereo). Cysteines were highlighted using the checkboxes in the top frame (out of view here). This makes the 6 Cys residues easy to spot for Chain A (boldface here at the lower left, green on the screen). Each Cys was clicked, which caused it to be rendered in ball and stick in the 3D image. Although the Cys residues appear positioned randomly in the sequence, the chain folds to create a geometrically regular 6-sulfur cage around the zinc ions. The Seq3D interface can also be used to select residues for display and colouring operations from the QuickViews menus (see *Figure 3*).

1.1 Rationale for Protein Explorer

The rationale for PE has been to make the power in Chime accessible to students, educators, and molecular biologists who are not experts in visualization software. The goal is to enable users to concentrate on molecular architecture, rather than software architecture.

Chime has enormous power, but by itself, most of the power in Chime is inaccessible to the majority of the people who could benefit from it, because of the technical knowledge required. PE may be thought of as a wrapper for Chime that strives to minimize the technical knowledge required, and to automate some commonly useful tasks that would require relatively heroic efforts to achieve in 'bare' Chime or RasMol (examples were listed above).

RasMol (2, 3), authored by Roger Sayle, is likely in use by millions of people. Despite its ease of use, relative to earlier software, we have observed that most college faculty are unable to use it effectively on an occasional basis. During 1997–99, we taught a series of NSF-supported 3-day workshops attended by nearly

QuickViews: 1d66.pdb

· SELECT	· DISPLAY	COLOR
1. SELECT ▼	2. DISPLAY ▼	3. COLOR ▼

SELECT	DISPLAY	COLOR
>HELP<	>HELP<	>HELP<
(Repeat)	(Repeat)	(Repeat)
All	Backbone	Structure
Chains	Trace	Chain
Protein	Cartoon	N->C Rainbow
Nucleic	Ball+Stick	Element (CPK)
Chain D	Stick	Polarity2
Chain E	Spacefill	Polarity3
Chain A	--------	Polarity5
Chain B	HBonds*	ACGTU
Residue	SSBonds*	ACGTUbb
Range	Dots*	Temperature
Solvent	Surface*	
Ligand	", Transp.	Black
Element	Contacts*	Blue
Hydrogen	*Hide*	Brown
Inverse	Cation-Pi	Cyan
	Salt Br.	CyanDark
Helices	Clicks	GrayLight
Strands	Center	Gray
Alpha C	Only	GrayDark
Backbone	Hide Sel. ▼	Green
Sidechain		GreenDark
Neutral		Magenta
Aromatic		MagentaDark
Hphobic		Orange
Polar		Purple
Charged		Red
Acidic		Violet
Basic		White
Cysteine		Yellow
Cystine		

Figure 3 QuickViews Menus. Exploration proceeds by selecting, then displaying (or hiding), then colouring groups of atoms. Here are shown the options in the current version of PE. Each operation automatically displays a detailed explanation (see example for DISPLAY, Contacts in *Figure 4*). Colour schemes automatically display a text key in matching colours.

100 college faculty from a dozen states in the Northeastern USA. It takes over an hour to teach a novice how to obtain an image as informative as the image that PE offers automatically as its FirstView (*Figure 1*). By the next meeting of the workshop (a week later), most participants could not reconstruct that procedure from memory, and even if given the image, some had difficulty interpreting it. With the further passage of time, only the most computer-adept remain able to use RasMol effectively unless they use it regularly. PE is designed to enable effective use on an occasional basis.

1.2 Evolution of molecular visualization freeware on personal computers

The first freeware that brought macromolecular visualization to personal computers was David Richardson's *MAGE* (5), released in 1992. MAGE's forte is *presentations* called 'kinemages', and over a thousand were created, some quite excellent (many of those by Jane Richardson). MAGE offers a number of powerful and unique features, but does not lend itself as well to self-directed exploration of a structure as does RasMol, and its images are less striking and rotate less quickly than those of RasMol and Chime.

RasMol was released by its author, Roger Sayle (2), in 1993. RasMol is a brilliantly implemented program. The speed with which it can rotate a spacefilled, depth-cued image of a large protein remains unsurpassed to this day. It achieves this speed, in part, by making a good compromise between image resolution and quality. Other visualization packages (see Section 1) typically have higher resolution images, and hence rotation is much slower. RasMol's speed of rotation made

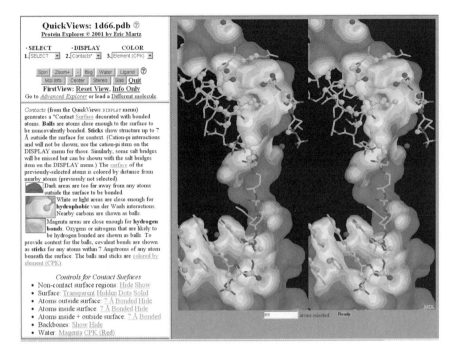

QuickViews: 1d66.pdb ⑦
Protein Explorer © 2001 by Eric Martz

·SELECT   ·DISPLAY   COLOR
1.[SELECT ▾] 2.[Contacts* ▾] 3.[Element (CPK) ▾]

[Spin] [Zoom+] [-] [Bkg] [Water] [Ligand] ⑦
[Mol Info] [Center] [Stereo] [Slab] Quit
FirstView: Reset View, Info Only
Go to Advanced Explorer or load a Different molecule.

Contacts (from the QuickViews DISPLAY menu)
generates a "Contact Surface decorated with bonded
atoms. **Balls** are atoms close enough to the surface to
be noncovalently bonded. **Sticks** show structure up to 7
Å outside the surface for context. (Cation-pi interactions
and will not be shown; use the cation-pi item on the
DISPLAY menu for those. Similarly, some salt bridges
will be missed but can be shown with the salt bridges
item on the DISPLAY menu.) The surface of the
previously-selected atoms is colored by distance from
nearby atoms (previously not selected).

Dark areas are too far away from any atoms
outside the surface to be bonded.
White or light areas are close enough for
hydrophobic van der Waals interactions.
Nearby carbons are shown as balls.
Magenta areas are close enough for **hydrogen
bonds**. Oxygens or nitrogens that are likely to
be hydrogen bonded are shown as balls. To
provide context for the balls, covalent bonds are shown
as **sticks** for any atoms within 7 Angstroms of any atom
beneath the surface. The balls and sticks are colored by
element (CPK).

Controls for Contact Surfaces
• Non-contact surface regions: Hide Show
• Surface: Transparent Hidden Dots Solid
• Atoms outside surface: 7 Å Bonded Hide
• Atoms inside surface: 7 Å Bonded Hide
• Atoms inside + outside surface: 7 Å Bonded
• Backbones: Show Hide
• Water: Magenta CPK (Red)

[89] atoms selected   Ready   MDL

Figure 4 Overview of protein–DNA interactions obtained in two clicks (convergent stereo). After loading 1d66 (see *Figure 1*) and going to QuickViews, two mouse clicks produced this image (SELECT, Chain A; DISPLAY, Contacts). The explanation and controls at the lower left appeared automatically. (To show more of this explanation, the frame containing the command entry and message box normally present at the lower left corner, as shown in *Figure 1*, was closed for this screenshot by dragging the frame boundary down.) Chain A is rendered as a surface, coloured by distance from nearby moieties as explained in the lower left frame. Four categories of interactions with the protein Chain A may be observed. (1) Water oxygens hydrogen bonded to the protein (isolated red balls). (2) Upper left: protein–protein interactions, largely hydrophobic van der Waals interactions (note preponderance of gray = carbon atoms [balls]). (3) Phosphorus atoms (orange balls) in phosphates along the right edge: electrostatic charge interactions that are nonspecific with regard to sequence. (4) Nitrogen-rich (blue balls) nucleotide bases at the lower left: sequence-specific recognition. A more detailed exploration of noncovalent bonds is available in the Noncovalent Bond Finder, an option in Advanced Explorer (not shown).

the use of stereoscopic viewing largely unnecessary in order to appreciate the 3-dimensional structure, and erased the former requirement for access to a super-computer in order to produce such images. RasMol excels at *self-directed exploration*, but is quite limited as a presentation tool.

The principal architect of *Chime* (Chemical MIME) was Tim Maffett in 1995–98, at MDL Information Systems. Chime 1 was released January, 1997. Chime 2, released in late 1998, was completed by Bryan van Vliet, Franklin Adler, Jean Holt, and others at MDL. Chime incorporates about 15,000 lines of C source code from RasMol, contributing the image rendering, rotation speed, and command

Figure 5 Enolase coloured to show conservation from a multiple protein sequence alignment. The sequence alignment, prepared outside of PE, includes enolase from eubacteria, yeast, drosophila, and humans. A 3D structure available for yeast enolase (4ENL) was loaded into PE. The sequence alignment was pasted into PE's MSA3D form (not shown), after which PE automatically generated the alignment listing shown, and coloured the molecule accordingly. Strikingly, not a single residue in the catalytic site (marked by the red sulfate ion that happens to be bound there) has mutated among the species aligned. The form (not shown) allows customization of the desired consensus level, the definitions of 'similar' amino acids, and the colours employed. The Chime command script that colours the molecule can be saved for use in a presentation. Thanks to Paul Stothard for contributing some of the javascript routines used in MSA3D, and to Garry Duncan for contributing the enolase alignment.

language control. (Notably, Roger Sayle placed no restrictions on derivatives of RasMol—see Section 1.3). Maffett added code to make Chime a browser plugin, to enable button-controlled scripting and other features to support presentations, support for movements and animations independent of hardware speed, rendering of solvent-accessible surfaces with colouring by electrostatic potential and lipophilicity (in collaboration with Michael Hartshorn), a more powerful menu, and numerous extensions to the command language. Maffett's work was also brilliant, and one of the best things he did was to retain nearly all of the capabilities in RasMol, including support for a command-line interface.

Chime, in collaboration with the browser, excels at *presentations*. On the other hand, its ability to support *self-directed exploration* remained largely inaccessible until the release of *Protein Explorer 1.0* in *1999*, which provided a slot for command-line entry, reporting of messages from Chime, and the ability to load molecules of the user's choice, all within the web browser.

The logical next step is to build support for presentations within Protein Explorer, enabling the viewer of a presentation or tutorial to move seamlessly between images in a presentation and self-directed exploration. At the time of this writing, this support is under development (see Section 6). A further step also under development will enable researchers to present their results in PE. The first example is the ConSurf server (9; see WebLinks).

See the weblinks box for more about the history of macromolecular visualization, including some historic photographs.

1.3 Freeware, open source

Most of the tools discussed in this chapter are freeware, and most, with the exception of Chime, are open source. PE may be downloaded in toto for off-line use. When you download it, you get its source, which comprises over 35,000 lines of javascript and HTML. The freeware Chime plugin's source code is said to be over 100,000 lines of C++, but the source code is not public.

1.4 Platforms supported and system requirements

Chime is available only for *Windows and Macintosh*, which limits PE to those two platforms. Linux users may use Protein Explorer in a Windows subsystem supported by either Win4Lin or VMWare (see WebLinks). Those using SGI, Sun, or other workstations may use Protein Explorer in a remote Windows NT session supported by Citrix Metaframe (see Weblinks, and `molvis.sdsc.edu/protexpl/platform.htm`).

If you use PE on-line, the only disk space requirement is for a web browser. If you download PE for faster use, or use off-line, PE itself will use less than twenty megabytes of disk space.

1.5 Browser compatibility

Chime and PE work in either the Netscape Communicator browser (Windows or Macintosh), or in the Microsoft Internet Explorer browser (Windows only).

Unfortunately, Netscape 6 was released with numerous bugs that prevent opera-tion of most Chime websites. Unless and until Netscape corrects these problems, Netscape 4.x (e.g. 4.80) is the browser to use with PE. (Netscape has kept all old versions readily available.) PE checks your browser at startup and informs you of any compatibility issues, and how to correct them. These checks will be updated as new browser versions are released.

2 Finding Chime-based tutorials

PE is not necessarily the best place to start investigating a molecule. Perhaps the molecule or topic of interest has already been nicely covered in a dedicated tutorial on the web. Chime-based tutorials have been developed for over a hun-dred macromolecules. For technical reasons, there is no way to limit a search of the WWW to pages that use Chime. The best place to start looking for existing tutorials is at the *World Index of Molecular Visualization Resources* (WebLinks). This is a visitor-maintained system. Visitors to the website can enter new URL's with descriptions and cross-indexing terms. As soon as approved by the moderator, the new entries are immediately alphabetized into the title, subject, and author indices of the appropriate category. Visitors can also edit existing entries. There are categories for Chime-based or non-Chime-based resources, and for biochem-istry, organic chemistry, inorganic chemistry and inorganic crystals, K-12, sources of molecules, DeepView (see Section 10), and free on commercial software, among others. Of course, even if you find useful information on your molecule, you will likely learn a great deal more by exploring it on your own. Read on.

3 Finding molecules

3.1 Protein Explorer requires a PDB file

PE cannot display a molecule if given only the amino acid sequence. In order to display a molecule, PE requires an *atomic coordinate data file* that specifies the identity and position of each atom in three-dimensional space. These are often in the Protein Data Bank format, hence called '*PDB files*'. The Protein Data Bank (see WebLinks) is the single, internationally recognized repository of all published, experimentally determined macromolecular structures (4). Each structure pub-lished at the Protein Data Bank is assigned a unique, 4-character *PDB identification code*, the first character of which is always a numeral (the remaining characters may be either numerals or letters). Examples are '1d66' (yeast gal4 transcriptional regulator complexed to DNA, used in the tutorial for PE), '1hho' (oxyhemoglobin), '1bl8' (the potassium channel), '1aoi' (nucleosome), '1ffk' (archeabacterial large ribosomal subunit).

Given amino acid sequence alone, present theory is able to predict secondary structure only to about 70% reliability, and is not able to predict tertiary or quaternary structure reliably. If a sufficiently sequence-similar protein's structure has been experimentally determined, a 3D structure for the sequence can be homology modelled with fair reliability. Homology modelling can be done, for example, with SWISS MODEL and DeepView (see WebLinks, and molvis.sdsc.edu/protexpl/homolmod.htm), and the resulting atomic coordinate file may then be explored in PE.

Most PDB files are empirical results of X-ray crystallography; about 15% are empirical results from solution NMR, and a few come from other empirical methods (see *Nature of 3D Structural Data*, www.rcsb.org/pdb/experimental_methods.html). Some theoretical PDB files are published at the Protein Data Bank—*caveat emptor*.

The Protein Data Bank has over 20,000 PDB files. Many of these, however, represent the same proteins modified by mutation or the binding of different ligands. So the world database represents only a few thousand sequence-unrelated proteins. Only a fraction of these are human, so we have structures for only a few percent of the roughly 35,000 proteins estimated to be coded for by the human genome.

PDB files are not equally available for all types of proteins. Proteins that contain hydrophobic regions, such as integral membrane proteins, are difficult to crystallize and few cases have been solved.

3.2 Finding PDB files

There are numerous search interfaces to the database of the Protein Data Bank (see the Molecules category at the World Index, WebLinks). If you are having difficulty finding what you want, it is wise to try several of these, because each has unique capabilities and flaws.

A search interface called *PDB Lite* (WebLinks) was designed for those who search for PDB files only occasionally. This is not to be confused with SearchLite offered at the Protein Data Bank. PDB Lite differs from SearchLite in the following ways: Jargon is avoided or explained, and crystal technicalities are omitted. Synonyms are expanded automatically ('hemoglobin' finds 'haemoglobin'). A checkbox option finds only sequence non-redundant structures, useful in cases where there are too many hits. Findings are tabulated with aligned columns easier to scan. Entering a query that won't work produces a warning with an explanation of how to rephrase the query. Step-by-step instructions are offered for saving PDB files in Windows or Macintosh, so you can successfully open it in PE from your disk. The database is updated weekly with the latest releases, and mirror sites are available on several continents.

4 Using Protein Explorer

PE is designed to be self-explanatory. At its FrontDoor is a link to the built-in 1-Hour Tour which guides beginners through the main features. Here, we'll provide further practical guidance, but please bear in mind that PE is constantly being enhanced, so some of the details may change.

4.1 Using PE's One-Hour Tour

Go to *proteinexplorer.org* (the leading www is optional), and you will see PE's main page, called the FrontDoor. This is the page to *bookmark*. Because PE has a complicated start-up sequence involving frames, bookmarking it at later stages is unlikely to work. Click on the link to the *One-Hour Tour*. It is a good idea to print the One-Hour Tour. Follow the One-Hour Tour to get an overview of the most important capabilities of PE. The on-line One-Hour Tour will be updated as PE is enhanced, so we will not repeat it here. The remainder of this section will make the most sense if you are sitting at your computer actually using PE and following the One-Hour Tour. If you have difficulty getting PE to show you the molecule, click on the 'Troubleshooting' link on the FrontDoor.

4.2 FirstView and spinning

After your session has started, you'll see the molecule rotating slowly ('spinning') together with a description of the FirstView (*Figure 1*). The purpose of the FirstView page is to explain the starting image, which is automatically rendered and coloured to be maximally informative. The links on this page provide detailed background and explanations of terminology for novices. All images are clickable. You are ready to leave FirstView when you know what the 'chains' represent, how many chains are present in the molecule, whether they are protein or nucleic acid, whether they have disulfide bonds, approximately what portion of the water is visible, and whether any ligands are present.

Initially, the molecule is *spinning* to emphasize its three-dimensional structure. After about 30 s, a box will pop up advising you to stop spinning because it takes a lot of computing power and will greatly slow the responsiveness of your computer. It is a good idea to leave the molecule spinning only when gazing transfixed at its beauty. Stop the spinning before pressing on links or buttons and the responses will be much faster. You can always drag on the molecule with the mouse to rotate it, and you should take advantage of this often.

4.3 Window management

PE is a large, complicated program, and often it will open additional windows. If the window you want sometimes disappears, and you can't find it, this section will help.

Windows: Notice that there are two browser buttons on the Taskbar at the very bottom of your screen. Try clicking them alternately to bring one or the other window to the foreground. The other window is still there but is hidden behind the one in front. Click on the link *Form for Recording Observations* near the bottom of the FirstView frame. Notice that there are now three browser buttons on the Taskbar. Notice that the Back button is greyed out at the upper left of the *Form* window. That means this is a new window (it has no history of earlier documents to go back to). When you are finished with such a window, close it by clicking on the X button at the very upper right of the window. (The main PE window has no Back button because it is not useful there—simply close it when you are finished with the session.)

Macintosh: Window management is similar, but use Communicator menu at the top of the screen, or, using the View menu: select Show . . . Floating component bar'. A floating button bar pops up; click the 'Navigator' button to switch windows.

4.4 QuickViews menus

If you are on the FirstView page, click the 'Explore More!' link near the bottom to move forward to QuickViews. The QuickViews menu system is the heart of the user-friendly power of PE. The menus retain the paradigm (while expanding the choices) created by RasMol, namely, *select* first, then *display* what you have selected, finally apply a *colour* scheme (*Figure 3*). This 3-step sequence can be repeated over and over, sometimes to change the rendering and colouring of a moiety, sometimes to render and colour a different moiety, sometimes to hide a moiety, and so forth. For example, SELECT All, DISPLAY Spacefill, COLOUR Polarity2. Notice that each operation automatically displays specific help in the middle frame. *It is important to read this help—it contains a wealth of crucial information.*

If you request the same action from one of the pull-down menus twice in a row, the second time has no effect. This is a technical limitation that we have been unable to fix. Use the '(Repeat)' option at the top of each menu when needed.

Notice that the *number of atoms selected* is always displayed in a slot below the molecular graphic (*Figure 1*). This is helpful in gauging the effect of your selection choice.

Also below the molecular graphic is a *Ready/Busy* indicator. Wait until PE is ready before you initiate another operation. In general, it is a good idea not to rush PE—clicking buttons rapidly may cause it to *freeze or crash*, despite the effort that has been made to avoid this. If PE stays Busy for a prolonged period, you can try clicking the Force Ready link below the message box (lower left frame). If that fails, the best thing to do is close the PE window and start a new session. Because PE, Chime, and the web browser are each large programs under constant development, there are inevitable bugs and some inherent instability. As with all complex applications, if strange behaviour persists, sometimes all browser windows must be closed (Quit on Macintosh) so that the browser can be reopened. Rarely, rebooting is required.

If you scroll down the QuickViews panel, you find the QuickViews Plus Options. These add an enormous amount of flexibility and power.

4.5 Secondary structure, hydrophobic core, sequence to 3D, quaternary structure, cation–pi interactions, salt bridges, etc.

The One-Hour Tour will show you how to see the amino and carboxy termini, secondary structure, the distribution of hydrophobic residues, net charge, and the noncovalent bonds between, for example, ligand and protein. In the 'External Resources Window', you will learn how to view the specific oligomer of any molecule, and to find out if contacts between chains are crystallographic artifacts. One-click QuickViews operations highlight cation–pi interactions (6) and salt bridges.

In later sections of the One-Hour Tour document ('Beyond the One-Hour Tour') you will learn how to view the sequences of the chains, clicking on residues in the sequence to identify their positions in the 3D image. This latter capability involves opening a window called 'Seq3D' (*Figure 2*), and enables you to select residues or ranges in the sequence for later changes in rendering or colouring using the QuickViews menus. For example, the Seq3D interface makes it easy to highlight residues known to be critical in an enzyme's catalytic site.

4.6 Help, Index, Glossary

Even after you have spent some time with the QuickViews menus, you will often find that you don't know how best to do what you want with the molecule. When this happens, consult the Help/Index/Glossary. It is linked to the FrontDoor (under About PE), and to a green circled question mark near the title 'QuickViews' (or the titles of other control panels in PE).

4.7 Loading your own molecule

There are several ways to get a molecule of interest loaded into PE.

- A number of search interfaces offer a *direct link* to PE after you have selected a molecule (see list on PE's FrontDoor).
- If you know the PDB ID code you can enter it in the slot on the FrontDoor, and PE will fetch the PDB file via Internet.
- If you have a previously downloaded PDB file on your disk, you can load it into PE by entering 'Bare' PE (use the link to 'Bare' PE on the FrontDoor). Since no molecule was pre-specified, PE opens with a Load Molecule screen. Press the Browse button and follow the instructions on the screen.
- The easiest and most reliable way to download a PDB file is with PDB Lite (WebLinks), which gives step by step instructions. You can also save any molecule directly from Chime while you can see it. (For instructions, open PE's Help/Index/Glossary with the link under About PE on the FrontDoor, and look up 'Saving the Molecule of Your Choice'.) Once you have the PDB file on your disk, the easiest way to load it is by entering 'Bare' PE (use the link to 'Bare' PE on the FrontDoor). Since no molecule was pre-specified, PE opens with a Load Molecule screen. Press the Browse button and follow the instructions on the screen.
- Home-made hyperlinks to PE can pre-specify the molecules to be displayed. There are a number of 'Quick-Start' links on the FrontDoor that illustrate this.

It is easy to make such hyperlinks on your own website; instructions are given on the FrontDoor page. The PDB file can come directly from the Protein Data Bank, or from any web server. If the former, PE allows you to set a preferred mirror site—this setting is a preference specific to your computer that will survive between sessions. Initially, PE will obtain PDB files from the San Diego headquarters of the Protein Data Bank (www.pdb.org). To change this, at the QuickViews screen, click on <u>different molecule</u>, then at the Load Molecule page, click on the blue question mark near the PDB ID code slot.

4.8 Advanced Explorer

Below the button cluster in the QuickViews frame is a link to Advanced Explorer. Here, one link goes to the Noncovalent Bond Finder (NCBF), a powerful and very easy way to get a detailed series of clear images of the noncovalent bonds surrounding any selected moiety. You can restrict the series to a subset of bonds, for example, hydrophobic interactions only, further to simplify the image. (NCBF is currently a separate system that was finished before PE was begun; someday we hope it will be properly integrated into PE.) Advanced Explorer also offers links to forms that customize contact surface displays, and customization to include ligands in the identification of cation–pi interactions and salt bridges. Finally, it offers forms into which multiple protein sequence alignments can be pasted, automatically colouring the 3D protein by conservation/mutation at the push of a button (MSA3D, *Figure 5*). Each of these advanced features has its own built-in tutorial.

4.9 Command language and aliases

The menus and buttons in PE all work through javascript (a programming language extension of the web browser, not to be confused with Java). PE's javascript generates sequences of commands and sends them to Chime. These commands control the appearance of the molecular graphic. As you use the QuickViews menus and buttons, you will notice 'messages' appearing in the box at the lower left, just below the slot marked 'Commands may be entered here'. (On Windows but not Macintosh, the most recent messages appear at the top.) Each command sent to Chime is displayed in this message box. Chime also responds to some commands with messages such as counts of atoms selected.

 If you are interested in learning the command language, a good way to do so is to watch the commands generated by PE, and then try entering some of them yourself. Knowledge of the command language will enable you to do some things more efficiently, but it is not worth learning unless you plan to use PE often. There is a more detailed introduction to the command language in PE's Tutorial.

 PE has a set of pre-defined command aliases. Aliases are abbreviations that make commands much easier to type, and reduce errors due to mistyping long commands. For example, 'bs' is an alias for 'Ball and stick'. When 'bs' is entered in the command slot, it is expanded to a two-command sequence (try it!). The list of pre-defined aliases is displayed with the Aliases link below the message box. Aliases are defined in a plain text file, so you can easily add new ones, or change, or delete existing ones.

4.10 Preferences

Below the message box is a link to Preferences. These settings are specific to your computer, and survive between sessions. For example, you can specify that water not be shown initially, and that the molecule not start out spinning. If you check the Expert preference, PE will bypass the FirstView explanation, and it will omit some help intended for beginners.

4.11 Off-line use and project folders

PE may be freely downloaded. This enables it to run off-line. Advanced users interested in writing scripts for presentations will be interested in PE's Project Folder capability. ('Folder' is used here as synonymous with 'directory'.) Below the message box is a link to set a project folder. This works only when PE is running from a downloaded copy. Once a project folder is set, PDB and script filenames used in manually entered 'load' and 'script' commands will be expected to be in the project folder.

5 Printing molecular images from Chime

Images on a computer screen have much lower resolution than do most printed images. Chime makes a compromise between image quality and the important ability to rotate the image relatively smoothly on ordinary personal computers. Chime has no ability to export a higher-than-screen resolution image for printing. (RasMol can export certain images in vector postscript or PovRay format; for more information, try asking on the Molecular Visualization Freeware Email Discussion, see WebLinks.) For publication purposes, WebLab Viewer Lite (see below, and Weblinks) can make much better quality images than can PE (Chime), for the types of images it supports. If you plan to print a molecular image from PE, it is best to start the PE session on the highest resolution screen available to you, for example, 1600×1200 pixels, making the PE window as big as possible on that screen.

Once you have the desired high-resolution image on the screen, the next step is to capture it for printing. (Printing directly from PE is generally unsatisfactory.) Click on the **MDL** below the molecular graphic, and a menu will open: select Edit, Copy. The molecular image is now in the clipboard. You can paste it into any suitable program, such as Word, Powerpoint, or Adobe Photoshop. If you wish to put it on the web, it should be saved to a file in GIF, JPG, or PNG format (see Chapter 8).

6 Presenting molecular images on-line

Images from PE can be presented as still snapshots, or as rotating displays in Chime. Snapshots can be viewed in Word, PowerPoint, or any browser, even when Chime is not installed.

6.1 Still snapshots

See the previous section about printing, which explains how to make a compressed image file containing a snapshot of a portion of the computer screen. Such snapshots can be used in *Powerpoint^{tm} presentations*, or inserted into web pages (in GIF or JPEG format only). Chime cannot rotate molecules directly in Powerpoint. If you want to show a rotating image during a Powerpoint presentation, simply have PE running with the image you want in the background, and at the appropriate times in your presentation, switch windows. If you are satisfied with a non-rotating still image, the screen snapshot file can be displayed directly in Powerpoint.

6.2 Presentations in Chime

Presentations in Chime can be used in two ways: projected, for illustrating a *lecture*, or by an *individual* working directly on a computer. Ideally, the formats for these two modes would be different and optimized for each use. However, few Chime website authors have begun to implement dual modes (but see recent work by Reichsman or Driscoll under Biochemistry Tutorials in Chime, at the World Index—see WebLinks).

Over a hundred presentations in Chime treating specific molecules are on the web (see World Index under WebLinks). The earliest presentations were by David Marcey and Henry Rzepa. These established a *two-frame format*, one containing Chime, and the other a scrolling tutorial in text. The text contains buttons, each of which produces the appropriate image in Chime. This two-frame format sacrifices at least one third of the screen area to the text frame.

It may seem undesirable not to use nearly all of the screen for the molecular image. In 1996, E.M. spent considerable time developing a tutorial format that used nearly all of the screen for the molecule, leaving only a small area for a figure legend, colour key, and three control buttons (back, next, quit). With great anticipation, he watched this used in a lecture in front of 500 students. During the lecture, a student asked a question about an image several steps back. Because the controls did not make it easy for the lecturer to find that particular image quickly, the opportunity was lost. This experience convinced him that *one must sacrifice enough of the screen for controls to enable the lecturer to find any desired image easily*. Furthermore, the small area reserved for legends and colour keys was inadequate. So he returned to the original two-frame format.

To make the format more appropriate for use in both lecture and individual modes, he provided buttons that can *increase or decrease the font size at will* during the presentation. In large lecture halls, a large font enables people in the back of the room to read the legends and colour keys. Individuals can reduce the font size so as to see more of the text at once, getting a better sense of where they are.

In 1997, E.M. provided a free downloadable *template* to facilitate authoring two-frame presentations in Chime (see UMass Chime Resources at WebLinks). The template frees the author of having to learn any javascript, and provides heavily commented HTML, easy to customize with no prior knowledge of HTML. This template included the font size control mechanism, detailed instructions, and access on each page to some utilities. The utilities provide, for example, a button to change the default black background to a white one for printing, or to enable mouse clicks to identify atoms or measure distances.

Most of the Chime presentations on the web today require that the buttons be pressed in order. The command scripts that generate the images were implemented so that most depend on the previous one. This has the unfortunate consequence that *if one skips buttons, or backs up, the resulting image is usually incorrect*. This can be easily avoided if every script begins by loading the relevant molecule anew, and then renders and colours it as desired. Each button is then guaranteed to produce the desired image regardless of the path taken to the button.

Reloading the same molecule over and over for a series of buttons that display it in different ways may at first seem unacceptably inefficient. In fact, this is not a problem, with the exception of PDB files larger than several megabytes. Chime has this built-in intelligence: if a script requests loading the same molecule that is currently already loaded, it merely resets the orientation and rendering as though it had been reloaded, without actually reloading the coordinates (see the command 'set load check' at MDL's Chime support site, under Weblinks; the default is 'set load check true'). The ultimate test, of course, is how rapidly the image appears when the button is pressed. If there is a significant delay between pushing the button, and getting the image, unless the PDB file is very large, the problem is likely to be that the script was generated by Chime or RasMol. Both generate scripts that always begin by loading the molecule, but are unnecessarily long. Instructions for shortening scripts are at www.umass.edu/microbio/chime/shared/sshorten.htm. A script recorder that will solve some of these problems is under development for inclusion in a future release of PE.

E.M. has developed a set of recommendations for the design of presentations or tutorials in Chime (see WebLinks). These include:

- Buttons should give the correct images regardless of the order in which they are pushed.
- The conclusion to be drawn from each image must be clearly stated in the button label.
- The button label should be short (a sentence or two). Details about the image should be shown in a pop-up window, with the pop-up mechanism being turned off for lectures.

- Standard colour schemes should be used when applicable (see DRuMS under WebLinks).

- A colour key (text in the same colours as the molecule) should be near the button for every image.

- Avoid pure blue on a black background—it is almost invisible when projected.

- If the presentation is put on the web, all PDB files should be gzipped (see FAQ at the UMass Chime Resources site, under WebLinks). Gzipping is a type of compression that speeds up the arrival of the molecule by about 3.5-fold.

- For use in a lecture setting, animations should be short, and used mostly to show transitions.

Regarding the last point, some authors have exploited the powerful 'move' command in Chime to make *movies* that last over 15 s. One problem is that the viewer may not realize the movie is still running, and push the next button. There will be no response, and other buttons may be pushed rapidly, at best leading to a lack of confidence, and at worst leading to a crash. Long movies also make it hard for a lecturer to control. Rotation to show three-dimensionality is best left to the mouse, rather than being forced on the user with a time-consuming movie. However, many of these limitations can be overcome using sophisticated javascript tools. A busy/ready indicator can inform a user that a movie is loading or running, and repeated button clicks can be intercepted before they cause any disruption. Also, movies can be paused at any point to allow a lecturer to discuss fine points. Finally, extended movies, even movies that loop constantly, can be especially useful in a non-lecture setting, where individual users may be less likely to use the mouse to manipulate a structure.

Probably the most justifiable use of animations is to clarify the *transition* between two images (two buttons). For example, a global overview of a large protein may be followed by a magnified view of the substrate-binding pocket (possibly with some chains hidden). It may be difficult for the user to follow the relationship between the two images if the jump is sudden. A movie showing the overview moving closer, and the hidden chains fading away, makes the relationship much clearer. This type of transition movie has been used consistently and with great effectiveness in the Biochemistry in 3D site (see WebLinks).

Often, during presentations, one wishes to be able to digress from the images provided by the author, freely to explore questions that arise during a lecture or discussion. This led to the next logical step: building presentations.

6.3 Presentations in Protein Explorer

E.M. is currently developing a new template to support presentations within PE. The presentation will have the conventional two-frame format, with controls to change font size. However, at any image in the presentation, one has the option of moving seamlessly into PE in order to use its exploration capabilities on the structure currently loaded—then back into the presentation. Check the PE FrontDoor where this will be linked when it is released.

Presentations in PE will include support for pop-up detail windows that can be disabled when lecturing. In a prior independent implementation, these have been used very effectively in Biochemistry in 3D (see WebLinks). It will also include a script recorder (programmed by T.D.) to help in authoring presentations.

7 Aligning two or more molecules

It is often useful to align two molecules to compare their structures. There are several websites that search for 'structure neighbours', that is, chains or domains related to a query molecule by 3D structure, without reference to sequence. One that provides aligned single chains compatible with PE is the Combinatorial Extension site (see WebLinks). This site also allows you simply to specify two PDB ID codes (and a chain for each if it contains more than one), and it provides the structural alignment. Once you arrive at the alignment page, there is a direct link to display the alignment in PE.

The disadvantage of the previous method is that the alignment will be stripped of all atoms except the protein chains themselves. Missing will be all ligands and water, 'hetero' atoms in PDB format parlance. In order to align chains while preserving hetero atoms, the powerful freeware modelling program DeepView (formerly known as Swiss-PDB Viewer) is relatively easy to use and quite effective (see below, and WebLinks).

8 Movies of conformational changes

Examples of movies of conformational changes in proteins can be viewed at the Protein Morpher site (among the UMass Chime Resources, see WebLinks). Here is a movie of the dramatic expulsion of an N-terminal myristic acid when calcium binds to recoverin, changing the protein from a soluble conformation to an insoluble, membrane-anchored one. A close-up movie of calcium binding to a single EF-hand (showing side-chain reorientation) helps to explain the mechanics involved.

Chime provides two mechanisms to display a series of different molecular conformations as a movie. One is built-in (see the 'Animate' toggle on Chime's menu, accessed by clicking on the 'MDL' frank), and employs atomic coordinate files in XYZ format, which allows multiple models to be specified. Great examples of XYZ animations are those by Motyka, Lahti, and Lancashire that illustrate vibrations in infra-red spectra (see WebLinks). But XYZ format is not useful for proteins because it does not identify atoms beyond their element. In order to draw a backbone trace, Chime needs to know whether a carbon atom is an alpha-carbon, and which amino acid it belongs to (residue name and sequence number). There is no provision in the XYZ file format for this information. The PDB format provides the requisite information, and allows multiple models to be distinguished (originally for NMR results). This enables a second mechanism for animations, more suited to proteins, that has been exploited in PE.

PE can animate multiple-model PDB files and morphs. Click on the animated icon of an EF hand binding calcium, on PE's FrontDoor, to view examples and find out how they are constructed.

9 Getting around the limitations of Chime

No single program can provide all of the features that a molecular visualization project might require. While PE is powerful and extraordinarily flexible, like all Chime-based software it is naturally limited by the capabilities of the plug-in. Chime cannot load and independently manipulate more than one molecule at a time. It cannot align or clean structures, mutate, or change the conformation of amino acids, form or break bonds, or model a protein structure based on its primary sequence. You cannot use Chime to build your own molecules or perform energy minimizations, nor does Chime provide useful information about bond angles, chirality, charge, stereochemistry, structure validity, bond order, or steric clashes. Yet all of these can be very valuable features, even for the occasional user of molecular visualization.

To complement PE, and work around the inherent limitations of Chime, we recommend stocking your molecular visualization toolkit with two other programs. *DeepView Swiss-Pdb Viewer* (Section 10) provides excellent molecular modelling capabilities and is free. *WebLab ViewerPro* (Section 11) is not free, but is especially useful for building and manipulating small molecules, and is an excellent general resource. The free WebLab Viewer Lite is quite adequate for publication-quality images. See the first paragraph in this chapter for a list of other freeware molecular visualization and modelling packages that are beyond the scope of this chapter, but are well worth investigating.

10 DeepView Swiss-Pdb Viewer

DeepView Swiss-Pdb Viewer is a powerful molecular modelling tool for the desktop PC. (It was originally called Swiss Pdb-Viewer. Recently, the name DeepView was added, and may eventually replace the older name Swiss Pdb-Viewer.) It was developed by Nicolas Guex, Torsten Schwede, Manuel Peitsch, and others with support from GlaxoWellcome (7). Swiss Pdb-Viewer is free, works on multiple platforms, and has excellent Web-based support (see WebLinks). It is not as flexible as Chime in manipulating or customizing the display of a macromolecule, but it should be your tool of choice for aligning multiple structures, mutating amino acids, or threading a protein's primary sequence into a 3D template.

In this section, we will introduce the main features and limitations of Swiss Pdb-Viewer. For a more thorough explanation, including many detailed protocols, we recommend the excellent Swiss Pdb-Viewer tutorials created by Gale Rhodes (consult the DeepView section of the World Index, see Weblinks). The program itself also offers excellent help at its Web site (see Weblinks). PE also contains documents that explain how to use DeepView for certain purposes (consult PE's Help/Index/Glossary).

10.1 System requirements

Swiss-Pdb Viewer runs on Windows, Macintosh, and Linux platforms. Generally, you will need about 10 MB disk space, although you can get away with 1 MB for the barest minimum installation on a Mac or PC. Macintosh users must have a PowerMac processor, 5 MB RAM, and a 256-colour display. Windows users must have Windows 95 or higher, OpenGL., and a 486 processor or better (a Pentium is recommended). A connection to the Internet is recommended for the molecular modelling features.

10.2 Homology modelling with Swiss-Pdb Viewer

The most powerful feature of Swiss-Pdb Viewer is its ability to model the 3D structure of a protein based on its primary sequence, using a technique called *homology modelling*. Homology modelling is a relatively easy and rapid method for deducing the structure of an unknown protein. It also turns out to be surprisingly accurate. While there are an astronomical number of possible protein sequences, there are only a limited number of energetically stable structures which a polypeptide can adopt (*protein folds*). In addition, protein structure tends to be *conserved*, with similar structures carrying out similar functions. (This is not always the case, of course, but it helps to increase the statistical accuracy of homology modelling.)

Swiss Pdb-Viewer boasts a direct connection to *Swiss-Model*, a comprehensive protein structure database maintained by GlaxoWellcome. From within Swiss-Pdb Viewer, you can load a file that contains the primary sequence of a protein of interest. The program will contact the Swiss-Model server (over your existing Internet connection) and search the database for proteins with similar sequences. Using the results, the program constructs a plausible 3D structure for your protein based on its similarity to known structures and displays it. You can tweak the new structure manually, even resubmit it for multiple rounds of modelling, then save it in any number of file formats (including PDB).

An understated but very powerful aspect of the relationship between Swiss-Pdb Viewer and Swiss-Model is its timeliness. As soon as new protein structures are solved and entered into the database, they are available to your local copy of Swiss-Pdb Viewer for use in modelling unknown proteins. No software updates are required to give you immediate access to the most recent scientific data. Swiss-Model and other very useful databases can be accessed directly on the Web as well.

One limitation of Swiss Pdb-Viewer is that it returns no information from the modelling process except the structure (e.g. what structures were homologous, how close was the fit, were there any regions excluded from the process, etc.). This is especially troubling if you submit a sequence that has little or no homology to any existing structure; in this case, the modelled structure would be all but useless. You can obtain this information by submitting the sequence to Swiss-Model directly via its Website, but this is more cumbersome than having a report feature built into Swiss Pdb-Viewer itself. For links to Swiss-Model and a beginner's introduction, look up 'homology modeling' in PE's Help/Index/Glossary.

10.3 Aligning multiple structures

Swiss-Pdb Viewer has the ability to load multiple structures and handle them independently. If the proteins are each a single chain with the same number of residues, Swiss-Pdb Viewer can align their structures to provide the best fit. This is especially useful when investigating different conformations of the same protein; for example, oxygenated versus deoxygenated myoglobin. A structural alignment emphasizes minute changes in conformation that lead to significant changes in protein function. The aligned structures can then be saved as a single merged structure file, which can be read by other molecular visualization software like PE. For details, in PE, look up 'alignment' in the Help/Index/Glossary.

Structure alignment is a good example of how molecular visualization programs can be used in concert to achieve a desirable goal. One common strategy for designing animations in Chime is to create a single file with multiple structures, each designated as a separate model, and alternating between them (see Section 8 above). Swiss-Pdb Viewer can be used to create files that contain multiple aligned conformations of a protein; subsequently, the transition between conformations can be animated in PE or Chime. For detailed methods, consult the Animations page in PE (click on the animated EF hand image at the upper left of the FrontDoor).

10.4 Mutating side chains

Changes to a single crucial amino acid or group of amino acids can have an enormous impact on protein function. Consequently, it would be valuable to be able to create such *point mutations* and observe their effect on the 3D structure. Swiss-Pdb Viewer allows you to change any amino acid into another—simply choose the 'mutate' option from the toolbar and click on the target residue, then choose the new amino acid from the contextual menu that appears. You can also evaluate different rotamers of a side chain, or allow the program to choose the lowest-energy rotamer. For step-by-step instructions, look up 'mutation' in PE's Help/Index/Glossary.

10.5 Adding hydrogens

Many macromolecular structures are solved by X-ray crystallography. Put simply, a crystal of the macromolecule is bombarded with X-rays, generating a pattern of electron scatter called an *electron density map* (the exact procedure for generating this map is exceedingly complex, and beyond the scope of this chapter). The 3D structure of the macromolecule is an interpretation of the electron density map.

In most cases, X-ray crystallography cannot resolve the exact position of hydrogen atoms in a macromolecule, so most 3D crystal structures do not include hydrogens. This is unfortunate because hydrogens participate in many interesting and important interactions at the atomic level.

Swiss Pdb-Viewer has the ability to add hydrogens to a 3D structure, using common modelling algorithms to calculate their positions. What's more, this feature is available via a single mouse click; all of the complex calculations occur behind the scenes. The result can be saved as a PDB file for examination in PE.

10.6 Crystallographic transformations

The crystal used to generate a 3D structure contains multiple copies of a macromolecule, packed together in close proximity. This packing can give rise to interactions between macromolecules that would not normally occur in nature. Thus it becomes important to assess the effect of such crystal contacts on the final 3D structure.

The standard PDB file format includes information that describes the unit crystal; you can use this information to reconstruct the crystal by manually entering the appropriate atom records. On the other hand, Swiss Pdb-Viewer will automatically reconstruct the unit cell with a single mouse click (Tools: Build Crystallographic Symmetry . . .), allowing you to focus on evaluating the possibility of crystallographically induced artifacts. For examples, in PE, look up 'Crystal Contacts'. There you will find a step-by-step explanation of how to construct PDB files containing all crystal contacts in Swiss Pdb-Viewer, and how to view the contacts effectively in PE.

10.7 Summary

Swiss Pdb-Viewer is an outstanding desktop molecular modelling tool. The features that we have presented here distinguish this program from related software, and make it a valuable addition to your molecular visualization toolkit. If you are interested in knowing what else Swiss Pdb-Viewer can do, and it can do a lot more, download the program and spend a few hours going through the tutorial. Also visit Gale Rhodes' Web site for more tutorial information (see WebLinks). It will be well worth your time.

11 WebLab Viewer

WebLab Viewer is available in both a free 'Lite' version, and a commercial 'Pro' version (see WebLinks) Developed by Molecular Simulations, Inc. (recently changed to Accelrys, Inc.), it is an excellent program that integrates molecular visualization and modelling tools. It is not quite as good as Chime at visualization, and not quite as good as Swiss Pdb-Viewer at modelling; however, it does an admirable job of combining the two aspects in a single program. WebLab ViewerPro makes good use of customizable toolbars and drag-and-drop technology to create a very intuitive interface. (Toolbars can be fully customized with any menu items, and either docked to the main window or made to float in convenient locations.) It also allows you to build your own small molecules and combine different structures in the same file, it has excellent output options for producing publication-quality images (*Figure 6*), and it is moderately scriptable.

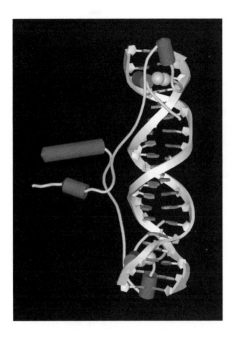

Figure 6 WebLab Viewer Lite rendering of Gal4 DNA-binding domain complexed to DNA (1D66, the same molecule as in *Figure 1*). Protein was displayed as 'Schematic' and DNA as 'Rings'.

Unfortunately, WebLab ViewerPro is not free, and the price tag (several hundred US dollars) is high enough to deter the occasional user. There is a free 'Lite' version which can produce higher quality images than either PE or Swiss Pdb-Viewer (e.g. for publication), but has severely limited capabilities, especially in modelling. If you can afford it, WebLab ViewerPro will be a worthwhile addition to your arsenal of molecular visualization tools; if not, WebLab Viewer Lite may be worthwhile for publication-quality graphics. In addition, if your work involves a lot of small molecule chemistry, and you want to work from your desktop computer, WebLab ViewerPro is a good choice.

11.1 System requirements

WebLab ViewerPro is limited to the Windows platform. It requires about 10 MB disk space, Windows 95/98/NT/2000, 32–64MB RAM, and a 32-bit colour display (recommended). WebLab ViewerLite, the free version, is available for both Windows (same requirements as the full version) and Macintosh (PowerPC with 8MB disk space and 32-MB RAM).

11.2 Loading, manipulating, and displaying molecules

One of the most helpful features of WebLab ViewerPro (and Lite) is its *clean and intuitive interface* for manipulating 3D structures. A single atom can be selected by clicking on it in the 3D structure; a group selected by double-clicking any atom in the group. Multiple atoms can be selected by dragging, or by using the *hierarchy window*, which shows all atoms arranged in a hierarchical tree according to group, chain, and molecule. Multiple windows with different structures can be open at the same time; alternatively, multiple structures can be opened in a single window using the *Copy From . . .* option.

Unlike many related programs, a selection in WebLab ViewerPro can be moved independently of the rest of the structure, even if the selection includes non-contiguous atoms, or atoms joined by covalent bonds to stationary groups. This allows easy creation of different rotamers, stereochemical conformations, cis and trans forms, and more. In addition, WebLab ViewerPro will display steric clashes (the *Monitor:Bump* option under the Tools menu).

WebLab ViewerPro (and Lite) provides tools for customizing most common display representations, including spacefill, ball and stick, scaled ball and stick, wireframe. Manipulating small molecules that are rendered as spacefill are no problem for WebLab ViewerPro; however, larger spacefilled structures slow down the response time of the program dramatically (in this area, PE/Chime and RasMol are still the undisputed champions).

WebLab ViewerPro offers a number of display options that are lacking from other molecular visualization software. For protein secondary structures, you can choose between *five different cartoon representations* (line ribbon, flat ribbon, solid ribbon, tube, or schematic), and *three different representations for nucleic acids* (ladders, rings, or arrows). Graphic quality during manipulation can be set to ultra-high (for fast processors) all the way down to low (for slow processors). There are options to turn on depth cueing, alternate between an orthographic and perspective projection, and change lighting colour and direction.

Especially useful is WebLab ViewerPro's ability to create different molecular surfaces. It will render two versions of a solvent-accessible surface (an approximate Soft surface, or a more exact Connolly surface) and let you alter the probe radius. It also offers a surface based on the van der Waals radii of the atoms. Surfaces can be coloured in a number of ways, including electrostatic potential, atomic charge or radius, or custom-coloured. WebLab Viewer Lite's rendering of surfaces coloured by molecular electrostatic potential is far superior to that of PE and Chime (see examples at molvis.sdsc.edu/protexpl/mep.htm).

Like all of WebLab ViewerPro's features, its many display options are readily available via menus, or they can be added to floating toolbars for immediate access. Most of these display options are available in the Lite version as well.

11.3 Information reporting

WebLab ViewerPro provides an extraordinary amount of *easily accessible information* about a 3D structure. Selected atoms are highlighted in yellow on the structure itself, making them immediately apparent. If an atom is selected, its identity is shown in the bottom right corner of the structure window (the status bar). If a group is selected, the status bar shows the group type, name and number (for example, AminoAcid: THR55).

Much more information about a selected atom appears in the Properties panel (Edit menu), including atom number, element, charge, hybridization, parent group, colour, and even the exact XYZ coordinates of the atom. Some of these properties can be changed directly, by clicking on the property name and entering the new value (see also Section 11.6).

A summary of all available information about a 3D structure can be accessed via WebLab ViewerPro's unique, spreadsheet-like data table (Window menu). No other popular molecular visualization program provides as much useful information in such a well-organized and intuitive interface. Each sheet of the table provides data about a different level of structure. The *cell* and *molecule* sheets show information about the structure(s) as a whole, including space group, molecular weight and formula, and number of atoms. The *atom* sheet lists every atom, with its parent group, element name, hybridization state, and charge. The *bond* sheet lists all bonds in the structure, including bond order, type, and length, and participating atoms. (Double-clicking on an atom or bond in the data table brings up its Properties panel.) This data table is not available in the Lite version.

11.4 Altering a structure

WebLab ViewerPro provides unparalleled access to the properties of a selected atom or bond. As mentioned in Section 11.3, some characteristics can be changed via the Properties panel in the Edit menu, including the name, number, colour, an atom's XYZ coordinates and partial charge, or a bond's length. More intrinsic properties are mutable via the *Modify menu*. Here, an atom can be changed into any other element via a clickable periodic table. Formal charge $(+1, 0, -1)$, hybridization (sp, sp2, sp3), and stereochemistry (S, R, relative, unknown) can be changed simply and easily. Finally, as in Swiss Pdb-Viewer (see Section 10), hydrogen atoms can be modelled onto a crystal structure in both WebLab ViewerPro and Lite.

If multiple structures are loaded into the same window (see Section 11.2), the *structures can be aligned* via an RMS fit against the first molecule. Structure alignments require the use of tethers, which are discussed in more detail in Section 11.6.

Making changes to a structure, especially using the mouse to drag atoms, often introduces non-standard geometries to bond angles and rotations. To resolve these problems, simply choose the *Clean Structure* tool from the Modify menu— standard geometries are restored with a single click. This feature can be active all the time (set using Options in the View menu), but this slows the response time of the program significantly; therefore, we recommend cleaning a molecule manually using the Modify menu.

11.5 Sketching small molecules

WebLab ViewerPro is noteworthy for its ability to *construct small molecules in 3D* quickly and easily. It operates in a similar fashion to the popular program ChemDraw. Toolbars supply a ready palette of pre-made bonds, atoms, and rings of all sizes, including chair and boat conformations of sugar rings. Atoms in the correct positions are automagically bonded together. Single, double, triple, and even aromatic bonds can be created with the click of the mouse. Finally, the geometry of the new structure can be optimized by cleaning (see Section 11.4).

WebLab ViewerPro's combination of design and rendering tools allows for an admittedly rough study of simple *docking interactions*; for example, how well a certain drug fits into the binding site of a potential target. Users interested in very precise interactions will require more powerful and expensive software that can perform energy minimization; however, the comparatively low price and minimal hardware requirements of WebLab ViewerPro may make it a useful intermediate tool.

A small molecule like a drug can be constructed in the same window as a target macromolecule like a protein, and the two structures can be manipulated independently, allowing you to approximate docking the drug into the protein binding site. Chemical changes to both drug and protein can be made on the spot; the effects of such changes can be examined by monitoring the resulting hydrogen bonds or bumps (steric clashes). Several rounds of careful tweaking can provide helpful information about preliminary drug design.

11.6 Tethering and aligning structures

WebLab ViewerPro does have the ability to align multiple structures that are loaded in the same window, as mentioned in Section 11.4. However, *to align macromolecules, we highly recommend using Swiss Pdb-Viewer*. The alignment process in Swiss Pdb-Viewer is much easier, more powerful, and gives much better results, despite being limited to aligning single chains. See Section 10.3 for more information on alignments in Swiss Pdb-Viewer.

In WebLab ViewerPro, structure alignment is based on *tethers*. Choose two atoms that you want to align (one from each structure), and select *Create Tether* from the Modify:Align Structures menu. Now select Align Structures to align the two molecules based on these two tethered atoms. You can define multiple tethers before aligning the structures; however, this does not guarantee that the alignment will be more accurate. For small molecules, this process is easy enough, but for macromolecules, it is all but useless.

11.7 Summary

WebLab ViewerPro is an excellent general resource for molecular visualization and modelling, especially of small molecules. It offers a number of unique display styles and several different high-quality output options. It provides detailed information about a structure, allows you to directly alter the chemical properties of atoms and bonds, and gives you the ability to sketch and clean small molecules quickly and easily.

The only serious drawback to including WebLab ViewerPro in your molecular toolkit is its fairly hefty price tag (several hundred US dollars) compared with PE and Swiss Pdb-Viewer (free). However, it's intuitive interface and unique features might make it worth the extra cost.

Acknowledgements

We are grateful to Roger Sayle, Tim Maffett, Jean Holt, and others who developed RasMol and Chime. E.M. appreciates support from the National Science Foundation Division of Undergraduate Education, and thanks Joel Sussman, Philip Bourne, and Walter Mangel for encouragement. Thanks to Paul Stothard for contributing portions of the MSA3D code employed in Protein Explorer, and to Frieda Reichsman for a thorough critique of this manuscript.

References

1. Marcey, David J. (2000). In *HMS Beagle: The BioMedNet Magazine* (`http://news.bmn.com/ hmsbeagle/90/reviews/insitu`), issue 90 (November 10).
2. Sayle, Roger A. and Milner-White, James E. (1995). *Trends Biochem. Sci.*, **20**, 374.
3. Bernstein, Herbert J. (2000). Recent changes to RasMol, recombining the variants. *Trends Biochem. Sci.*, **25**, 453.
4. Berman, H. M., Westbrook, J., Feng, Z., Gilliland, G., Bhat, T. N., Weissig, H., *et al.* (2000). The Protein Data Bank. *Nucleic Acids Res.*, **28**, 235.
5. Richardson, David C. and Richardson, Jane S. (1994). *Trends Biochem. Sci.*, **19**, 135.
6. Gallivan, Justin P. and Dougherty, Dennis A. (1999). *Proc. Natl. Acad. Sci. USA*, **96**, 9459.
7. Guex, N. and Peitsch, M. C. (1997). SWISS-MODEL and the Swiss-PdbViewer: an environment for comparative protein modelling. *Electrophoresis*, **18**, 2714. `http://www.expasy.ch/spdbv`
8. Glaser, F., Pupko, T., Paz, I., Bell, RE., Bechor-Shental, D., Martz, E., Ben-Tal, N. (2003). ConSurf: Identification of functional regions in proteins by surface-mapping of phylogenetic information. *Bioinformatics.* **19**(1), 163–4.
9. Martz, E. (2002). Protein Explorer: easy yet powerful macromolecular visualization. *Trends Biochem. Sci.*, **27**(2), 107–9.

Chapter 10

Virtual reality for biologists

Tomaz Amon

Center for Scientific Visualization, University of Ljubljana, Faculty of Electrical Engineering, Trzaska 25, 1000 Ljubljana, Slovenia.

1 Introduction

As the name implies, the Virtual Reality Modelling Language (VRML) is a computer language for describing three-dimensional structural models or 'worlds'. Much like HTML, it can be written with an ordinary text editor, then viewed with appropriate visualization software; this is the approach we use in this chapter to introduce the concepts.

This tutorial is written for molecular biologists who may not be primarily interested in producing virtual reality worlds with complex models and sophisticated interactivity, but instead just want to see and/or show their data in 3D. For many of these purposes, no powerful modelling software is needed; we just bring the set of 3D coordinates into 'virtual reality space' to produce a model.

VRML models are typically integrated into a web page. Originally, a VRML plugin was required to add viewing capability to a web browser, but more recently Java applets such as Shout3D have been developed that allow viewing 3D worlds without the need to install a plugin. Since it is designed for use on the Web, the language is intended to use both bandwidth and processing power economically; for example, a VRML file is commonly 100 times smaller than a video clip showing similar content.

Creators of more complex VRML worlds usually do not type the text directly, but instead use some dedicated VRML editing software (see *Table 1*). *Spazz3D* is a useful and inexpensive tool for modelling and animating interactive VRML worlds, and can directly export to Shout3D format. *VRML Pad* is a text editor dedicated to writing or editing VRML files. Typically it is used after a VRML world has been modelled (e.g. with Spazz3D) and some final improvements of the 'raw code' are desired.

2 VRML viewers

VRML was originally produced by Silicon Graphics and was further developed by the company Cosmo software. Its browser plugin Cosmo Player and the authoring

Table 1 Web resources

Java applets for 3D	
Shout3D	/www.shout3d.com
Tutorials and general reference	
Sandy Ressler's site	3dgraphics.about.com
Tomaz Amon's site	www.bioanim.com
VRML editors	
Spazz3D	www.spazz3d.com
VRML pad	www.parallelgraphics.com/products/vrmlpad
VRML plug-in	
Cortona	www.parallelgraphics.com/products/cortona
Other Technologies	
Adobe Atmosphere	www.adobe.com/products/atmosphere/main.html
Macromedia Director	www.macromedia.com/support/director/3d_index.html
Macromedia Flash	www.macromedia.com/support/flash/ts/documents/ 3d_approaches.htm
Discreet Plasma	www.discreet.com

Note: Older plugins, such as Cosmo Player, may not support the extrusion nodes used for many of the examples in this chapter.

tool Cosmo Worlds were excellent tools, and Cosmo Player is still widely used today as a 'standard' VRML plugin, though it is rather obsolete. The Cosmo software company is no longer producing updates, though further development of the technology is being done by other groups. The need for users to have VRML plugins installed in their web browser has been a significant drawback to more widespread use of 3D models on web pages. The original plugin (Cosmo Player) was more than 3 MB in size and it was not bundled with either Netscape or Internet Explorer, so it required extra effort on the user's part to download and install. Your website might have brilliant 3D content, but visitors would not be able to see it without the correct plug-in, and many might choose to skip the site rather than undertake the technical challenge of installing the necessary software.

Several avenues of development have improved this situation. First, plugins are much smaller and much more technically advanced (e.g. Cortona, *Table 1*). Second, using a different approach to reducing the technical burdens on potential users, Java applets have been developed that show virtual reality worlds without the need for a plugin. This solution works on Windows as well as on the Macintosh platform. Although these Java applets are not as powerful as a modern web3D plug-in, they are appropriate for simple models. Users who are suitably impressed and motivated may choose to download and install a plug-in to be able to enjoy more complex VR worlds.

Molecular biologists have many ways of showing molecules in sophisticated and modern ways. These topics are discussed elsewhere in this book (see Chapter 9); most of them require specific plug-ins that are specialized for viewing molecules. Here we shall demonstrate a more general-purpose approach to three-dimensional modelling that can be applied to a wide variety of datasets and other visualization needs. All you need is a web browser, a text editor, and the Shout 3D 2.0 package which you download for free from the web (see *Table 1*).

We shall create our VR models in VRML, which can be viewed using the Shout3D Java applet. As a Java applet, the Shout3D software is stored on the web server along with the VRML model, and downloaded automatically when it is needed. Alternatively, users who do have a plug-in can use it to see the same model files. The examples shown in this tutorial use only a tiny fraction of the possibilities that VRML offers. For more in-depth information, I recommend Sandy Ressler's site (see *Table 1*), which is an excellent resource for web3D tools and worlds.

3 Modelling a sphere

We start with the VRML model of a sphere. Type the text shown in part (a) of *Figure 1* into a text editor, and save it to disk. Part (b) of *Figure 1* shows the model as viewed using a VRML plug-in.

The first line "#VRML V2.0 utf8" is the VRML identification line. Please leave it as it is. The Transform{...} statement specifies the characteristics of the sphere. The colour is set in the line "diffuseColor 0 1 0" where the three parameters specify the levels of the red, green, and blue (RGB) colour components, respectively. Each colour parameter is a floating point number in the range from 0 to 1 (0–100% colour). As many HTML and image processing programs accept colour values as integers from 0 to 255, be careful to scale the colour parameters accordingly before porting them to VRML.

The "translation 0 1 0" line tells that the centre of the sphere is positioned in the VR space at the coordinates $x = 0, y = 1, z = 0$.

The code in *Figure 1(a)* is formatted to show some of the relationships between the objects. A transform object has a translation (position in space), one or more child objects, and optionally some other attributes, such as a scaling factor. The only child of this Transform is a spherical shape. The indentation is used merely to make these relationships more easily visible to humans; a VRML viewing program treats all groups of 'white space' (tabs, spaces, and line breaks) merely as separators between syntactically important words and symbols.

We shall now reformat the code to make it easier to add as many additional spheres as you like and set their individual colour, position, and size parameters. First let's format the code for this single sphere so that it fits in a single line and so the parameters we are interested in are separated by tabs (*Figure 2*). As mentioned above, changing the indentation, line breaks, and so forth does not affect how the VRML code works, but this will let us use some word processing tricks to edit our VRML model more easily.

Now replicate this line once for each sphere you need, and convert the text to a table in a word processor, as shown in *Figure 3*. Note that the colour, translation, and scale parameters are arranged as columns of numbers; this is because the

```
(a) #VRML V2.0 utf8
    Transform {
        children Shape {
            appearance Appearance {
                material Material {
                    diffuseColor 1 0 0
                }
            }
            geometry Sphere {
                radius 1.4
            }
        }
        translation 0 0 0
    }
```

(b)

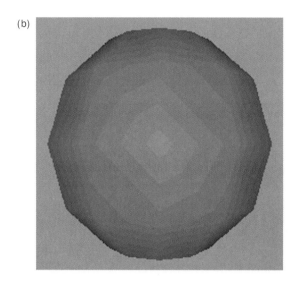

Figure 1 The VRML code for a sphere (a) and its corresponding model (b).

Transform·{·children·Shape·{·appearance·Appearance·{·material·
Material·{·diffuseColor→1 → 0 → 0 → }·}·geometry·Sphere·
{·radius → 1.4 → }·}·translation → 0 → 0 → 0 → }¶

Figure 2 The sphere code from *Figure 1(a)* converted to a single line with tabs separating the parameters. The VRML code was edited to replace the hard line breaks with spaces. Here the code was copied into Microsoft Word in order to make the white space characters visible; tabs are represented by arrows, the end of the line is shown by a paragraph mark, and spaces are represented by small dots at mid-height in the line of text. To view the VRML model, this code must be saved as plain (ASCII) text. Note that the three lines in the figure are 'soft wrapped'; there is only a single 'hard' line break, represented by the paragraph mark at the end.

194

Transform { children Shape { appearance Appearance { material Material { diffuseColor	1.00	0.00	0.00	} } geometry Sphere { radius	1.40	} } translation	0.00	0.00	0.00	}
Transform { children Shape { appearance Appearance { material Material { diffuseColor	1.00	1.00	1.00	} } geometry Sphere { radius	1.10	} } translation	0.97	0.00	0.00	}
Transform { children Shape { appearance Appearance { material Material { diffuseColor	1.00	1.00	1.00	} } geometry Sphere { radius	1.10	} } translation	−0.24	0.94	0.00	}

Figure 3 VRML pasted into a table where we can easily replace the colour, radius, and translation parameters. The text from *Figure 2* was copied into Microsoft Excel. As with most spreadsheet programs, pasting tab-separated text elements causes them to be automatically placed into columns. Two additional copies were made of this row of the spreadsheet, and the numbers for the colour parameters (red, green, and blue), the radius parameter, and the *x*, *y*, and *z* values for the translation parameters were set as desired. The numbers were formatted to two decimal places, then the data was pasted into a table in Microsoft Word to create this figure.

(a)
```
#VRML V2.0
Transform  ffuseColor  1.00  0.00  0.00 }  ius 1.40 }} translation 0.00  0.00  0.00 }
Transfor  diffuseColor  1.00  1.00  1.00 }  dius 1.10 }} translation 0.97  0.00  0.00 }
Transf   diffuseColor  1.00  1.00  1.00   radius 1.10 }} translation −0.24 0.94  0.00 }
```

(b)

Figure 4 A VRML model of a water molecule consisting of three spheres. (a) A part of the source code shown using one line per sphere; parts of the code have been deleted so that the parameters of each sphere can be seen in the figure. (b) The code of part 'a' viewed as a 3D model.

tabs we inserted around the parameter values cause a jump to the next column in the table, and the hard line breaks cause a jump to the next row.

Edit the numbers in the table to set the appropriate parameters for the individual spheres, then copy the text back to the VRML text file under the first line "#VRML V2.0 utf8", and the simple multiple-sphere model is complete (*Figure 4*).

What we have obtained is a molecule-like virtual reality model; you may recognize it as water. The numbers in the columns of *Figure 4(a)* come from a general chemistry text (1): oxygen has a Van der Waals radius of 1.4 Å, and is coloured red in the CPK scheme; hydrogen atoms are white with a radius of 1.1 Å; the O–H bond distance in water is 0.97 Å, and the H–O–H angle is 104.5°. The X and Y displacements of the second hydrogen are calculated from `r cos(θ)` and `r sin(θ)`, respectively. As in a dedicated molecule viewer, this molecular model can be rotated to view it from different angles.

The size of this file is only 531 bytes, so it downloads in less than a second at an average modem speed.

In addition to spheres, simple geometric shapes supported in VRML include the box, cone, and cylinder nodes. Ribbon models, commonly used to show molecular structure, are modelled in the VRML with the 'extrusion node'. It is almost as simple as the sphere node, as we see in *Figure 5*.

The code `"crossSection [0 0.87, 1 -0.87, -1 -0.87, 0 0.87]"` defines the 2 dimensional cross section (with *x* and *y* coordinates) of the shape to be extruded. It is a triangle in this example; later we shall use a polygon to approximate a circle. You can use almost any polygonal shape you want for the cross section. Note that the first and the last *x*-, *y*- coordinate sets (`"0 0.87"`) are the same, so that the cross section shape is 'closed'. The code `"spine [ 0 -1 0, 0 1 0 ]"` sets the *x*, *y*, and *z* coordinates of points along the spine of the extrusion in 3D space. In this case we have only two points, one with the *y* value −1 and the other with `y=1`.

As a single extrusion can represent quite complicated shapes, we do not need to replicate the 'Transform' nodes as in the previous example with spheres. We simply define a cross section, and specify a set of spine coordinates we would like to trace through space. As a further refinement, we can specify a different scaling factor for the *x* and *y* dimensions of the cross section to be applied at each spine point.

To illustrate the power of these extruded shapes, we will generate a fairly sophisticated 3D spiral. This shape is used to model the human cochlea, as described in the section 'Creating more complex virtual reality worlds' below. It is illustrated in *Figure 6*. The cross section is approximately circular, being defined as a polygon connecting 16 points around the circumference of a circle. The cochlear model uses 79 spine points evenly spaced along approximately two and a half turns of the spiral. The radius of the spiral increases by a fixed amount at each point, and the height along the Z axis increases. Note that the cross section is small at the more tightly curved end of the spiral, and gradually becomes larger at the other end; this is accomplished by setting a scale factor for each spine point.

Numerical calculations for such shapes are easily carried out with a spreadsheet program such as Microsoft Excel. To determine the *x*, *y*, and *z* coordinates of the 79 spine points, we first create a column in the spreadsheet to represent the angle θ around the axis of the spiral. Starting at 0, we add 0.2 radians at each step; the 79th step brings us to 15.6 radians, or about two and a half turns.

```
(a)  #VRML V2.0 utf8

     Transform {
       children  Shape {
         appearance  Appearance {
           material  Material {
           }
         }
         geometry Extrusion {
           crossSection  [ 0 0.87,
                    1 -0.87,
                   -1 -0.87,
                    0 0.87]
           spine [ 0 -1 0,
                   0 1 0 ]
           scale [ 1 1, 1 1 ]
           orientation  [ 0 0 1  0,  0 0 1  0 ]
           solid FALSE
         }
       }
       translation   0 1 0
     }
```

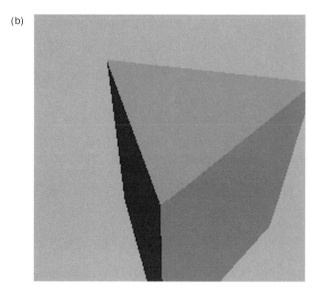

(b)

Figure 5 VRML extrusion in its simplest form, source code (a) and model (b).

In a parallel column, we calculate the radius r of the spiral for that point. The radius begins at 0.2, and increases by 0.05 at each step, reaching a final value of 4.1 on the 79th step. Now the x and y coordinates are calculated in additional columns: $x = r \cos(\theta)$ and $y = r \sin(\theta)$. The z values simply increase in constant steps of 0.1, ranging from 0 to 7.8. The scale factors increase linearly in steps of 0.01 from 1.00 to 1.78. There are two scale factors, one for the X dimension of the cross section, and one for the Y dimension. Since the cross section of our model changes proportionately, these X and Y scale factor numbers are the same for each point. Once the numbers have been calculated in the spreadsheet, the x, y, and z columns along with the scale factor columns can simply be copied

(a)
```
#VRML V2.0 utf8

Transform {
    children [ Shape {
            appearance  Appearance {
                material Material {
                    diffuseColor 0.5 0.5 0.5
                }
            }
            geometry Extrusion {
                beginCap TRUE endCap TRUE solid FALSE creaseAngle 1.43
                crossSection    [
                    1 0,  0.92 0.38,  0.71 0.71,  0.38 0.92,
                    0.00 1.00,  -0.38 0.92,  -0.71 0.71,  -0.92 0.38,
                    -1.00 0.00,  -0.92 -0.38,  -0.71  -0.71,  -0.38 -0.92,
                    0.00 -1.00,  0.38 -0.92,  0.71 -0.71,  0.92 -0.38,
                    1.00 0.00
                ]
                spine   [
                    0.20      0.00      0.0 ,
                    0.25      0.05      0.1
                    0.28
                                      -4.23      7.7 ,
                    -4.08      0.44      7.8
                ]
                scale   [
                    1.00      1.00
                    1.01      1.01          ,
                    1.02
                                      4.77      ,
                    1.78      1.78
                ]
            }
        } ]
    translation 0 0 0
    rotation 1 0 0 1.87
    scale 1 1 1
}
```

(b)

Figure 6 A VRML extrusion showing a three-dimensional spiral. (a) selected parts of the VRML code and (b) model.

198

and pasted into the appropriate positions in the VRML text file. In the default orientation, the XY plane is parallel to the face of the computer monitor, and the Z axis represents depth. Our model has been turned slightly more than a quarter turn around the X axis, so that the Z axis points more or less down. This is done by setting the `rotation` parameters; the first three numbers represent a vector to serve as an axis of rotation (the X axis in this case), and the fourth number is the rotation angle (in radians).

A simple example using an extrusion node to trace the backbone of a protein structure and show it in proposed relationship to the plane of the cell membrane is given in Chapter 11. While dedicated molecular viewing software offers many desirable features (such as dynamically changing representations and residue identification), VRML models offer the possibility of being combined with non-molecular models in a single environment. Sophisticated VRML models of molecular structures can be generated at the Protein Data Bank website (see Chapter 9) simply by choosing the VRML option from the View Structure page.

We have seen how to use two of the basic shapes supported by VRML, spheres and extrusions. Other shapes, such as boxes, cones, and cylinders are used very similarly to spheres. VRML also supports an 'IndexedFaceSet' node that can be used to create arbitrary 3D shapes. More information can be found in the on-line references. All of these shapes, and collections of shapes, can be positioned within Transform nodes, as we have done with these examples. Many shapes and 3D structures can be combined into 'worlds'; some more complex worlds are described below.

3.1 Creating HTML documents containing these worlds

VRML worlds typically reside in text files ending with the extension `".wrl"`. Viewers whose web browsers have VRML plug-ins installed can open these files directly, but web3D worlds can also be embedded in HTML documents. Here we describe how to place 3D content on the web using the Shout3D Java applet.

3.2 Downloading the software and organizing the folders

We will describe how to set up a website to use the freeware version of the Shout3D applet. You do not need to do this if you are using a plug-in, but as described earlier, it makes life easier for potential visitors to your website, because then they will not be required to install a plug-in. In the freeware version of the Shout3D applet (*Figure 7*), there is a black band in the lower part of the window with the Shout3D logo. This is a link to the company that produces this software. This logo does not appear in the commercial version.

First visit the site (see *Table 1*) and download the free Shout3D environment, which you can install on your server. We describe installation under Windows, but the Java code works on Unix and Macintosh as well. The Shout3D package is by default installed in the `C:\Shout3d` folder. Read the instructions and the Shout3D tutorial to get familiar with the software package. To establish the environment for your web3D site you first have to create a folder where your project resides. In our case this is the folder 'VRMLtutorial'. It can be anywhere

TOMAZ AMON

```
<applet MAYSCRIPT align=LEFT valign=MIDDLE
    name="Shout3D" codebase="../codebase"
    code="applets/ExamineApplet.class"
    archive="shout3dClasses.zip"
    width="242" height="276">
    <param name="src" value="models/extrusion5.wrl">
    <param name="backgroundColorR" value="0.7">
    <param name="backgroundColorG" value="0.7">
    <param name="backgroundColorB" value="0.8">
    <param name="headlightOn" value="true">
</applet>
```

Figure 7 The HTML code needed to embed a Shout3D world in a web page. The highlighted portions of the text are most often adapted to suit the individual needs. Please see text for a detailed explanation.

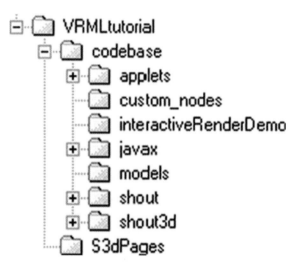

Figure 8 An example setup of Shout3D directories inside your project folder (in this case 'VRMLtutorial'). The folder 'models' contains the web3D models, the folder 'S3dPages' has the HTML files pointing to the models. The folder 'applets' contains applets for displaying your models on the web in several different modes.

on your disk. Then copy the folder 'codebase' from the C:\Shout3d folder to the VRMLtutorial folder. You store your VRML models in the 'models' folder within the codebase folder. Many sample files are included with Shout3D. You can erase them if you wish to have only your data on your site. Finally place your HTML files pointing to the web3D worlds in the appropriate folders (directions for putting the appropriate applet tags in a web page are given below). I usually store them in a folder like the 'S3dPages' lying inside the VRMLtutorial folder. The setup should look like the folder tree in *Figure 8*. The directory structure is important so that the applet can find the software components it needs. If the directories are not arranged properly, errors will result.

200

3.3 Adding the applet tag

The applet tag is a bit of HTML code that controls how your 3D world will be displayed on the web page. It points to a Shout3D applet and to one of the VRML worlds we have just created. Samples of HTML documents including such applet tags can be seen in the 'S3dPages' tutorial (*Table 1*). The code you need to write is very simple (*Figure 7*). Adapt the VRML file name (here 'extrusions5.wrl'), and store the VRML files in the codebase/models folder. The width and height parameters set the dimensions of the window that displays your VRML world on the web page. Larger windows require more computer effort to refresh when the user manipulates the content, so it is the best to keep it a reasonable size. The particular applet specified (such as 'ExamineApplet.class') controls how your VRML world is presented. The 'Examine' applet allows you to rotate and zoom your model—this is probably a default for molecular biologists. Alternatively, the 'WalkApplet.class' lets you walk through your models. Look in the folder codebase/applets to see all the viewing applets that are available. Finally you set the background colour with the parameters for red, green, and blue (RGB). You can even set a panoramic background if using one of the panorama applets.

3.4 Creating more complex virtual reality worlds

This brief VRML tutorial has shown only a tiny glimpse of the possibilities offered by a web3D software environment. Typically in such a VR world you see a scene in virtual space containing objects composed of collections of cones, spheres, or more complex 3D shapes, rather than flat painted images. In fact, one could potentially model and interactively animate almost any object. However, limitations arise when the computer renders this world on the screen. If the computer requires many seconds (or minutes, or hours) for a single frame, the user experience may not feel much like 'virtual reality' (at least 10 frames per second are needed for a web3D world to flow reasonably smoothly). Complex animations in the cinema involve recording VR scenes on film or video so they can be played back at a fast enough rate to simulate continuous motion. For web applications running on standard desktop computers the VR worlds have to be simplified so that they can be rendered on the screen in 'real-time'. Fortunately, simplification is often an advantage for educational purposes, since this can make structures and processes more understandable. Although the VR world is simplified enough to be stored in a relatively small file (typically a VRML file is less than 100 kilobases), its interactivity helps you become immersed in its space. You are repesented there by an 'avatar', from whose perspective you observe the environment. Since the software knows where your avatar is located in the VR world, it can interact with you. You can press buttons to elicit some behaviour or to trigger a hyperlink that brings you to another web page, like in any classical hypertext document. The target of the link may be anything your browser can display, such as a web page, another 3D model, etc.

A 'level-of-detail' (LOD) function sets on the scene more detailed versions of objects as you get nearer and simplified versions if you travel away. More

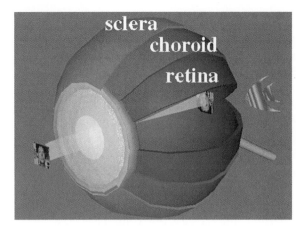

Figure 9 VRML model of the human eye. The structures are arranged as a LOD (level of detail) grouping so that when the user approaches the eye, its structure 'opens' and the appropriate functions of the eye are demonstrated (changing of the iris aperture). The fish represents a particular viewpoint from which the model can be examined.

generally, the LOD function can be used to trigger changes in how the scene is displayed, depending on how near the viewer is. This technique has been used, for example, in the structure of the human eye described in *Figure 9*. As the user approaches the model of the eye, it 'opens' to reveal the internal structure.

Of course, objects in a web3D world can also be animated. This is done by setting the important points of the animation (the keyframes) within a VRML 'Position Interpolator'; the animated objects move smoothly from the first to the last keyframe. The animation can be triggered by pressing a button or simply by the user entering a certain region of the virtual space.

Since the VR world provides primarily visual information, lighting is very important. By default in most viewers your avatar has a 'headlight' when traveling around, and other lights of different intensities and colours can also be specified. However, too many lights may slow down the rendering speed.

Sound can also be used in a VR world. You position a 'loudspeaker' in the space and define how much the sound attenuates with distance. When the user goes away from the sound source, the sound becomes quieter, diminishing completely at a given distance. For example, if the user clicks with the mouse on the tympanum in the model of the human ear (*Figure 10*), the tympanum and its associated ear bones move and a sound is heard over the computer speakers.

Applying a background panorama can add another element of realism. Panoramic images are typically stiched together from several pictures with an image stiching program. This panorama scene is then applied as the background of the VR world. Up to 6 background pictures (4 side pictures as well as top and bottom ones) can be used to create a landscape panorama behind your scene. The pictures need to be stiched together with appropriate software (*Table 1*: Sandy Ressler's site). As an alternative to using images, the VRML 'Background' node

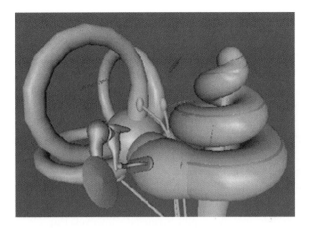

Figure 10 VRML representation of the structure of the human hearing organ. Clicking on various objects causes them to become animated to show a function in the ear, or to disappear to show the structures lying underneath.

allows you to specify background colours with optional transitions, for example, a range of colours from blue to orange might be used to simulate sunrise.

A 'texture' is an image that 'sticks' to the surface of your model, allowing you to adorn your models with nice pictures or patterns. However, these images significantly increase the size of your VRML files, and make your objects move or rotate more slowly and less smoothly. This is not the problem of VRML itself, but the technical limit of slower computers like a Pentium II PC. On my Silicon Graphics workstation, VRML worlds with textures run very smoothly, and fast Pentium IV and PowerPC computers with good 3D graphics accelerator cards go a long way towards solving this problem, too. To keep your virtual world simple enough to fit in a small file, load quickly, and run smoothly on an average computer, you should primarily use models painted with colours or simple textures. A world of about 60 KB is capable of showing many reasonably intricate models with interesting user interactions.

4 Alternative web3D technologies

Following is a short overview of popular web3D technologies currently in use in addition to VRML. Since these technologies change very rapidly, please look for updates on the web like the general reference sites of *Table 1*.

4.1 X3D

X3D is intended to be the successor of VRML. It combines the functionality of VRML97 with the data structuring syntax of XML. This approach provides extensibility and many other useful features. It has been developed as an open standard largely by Don Brutzman's group (*Table 1*: Sandy Ressler's site), which has also produced an immense database of X3D examples available on the web.

4.2 Shout3D

One of the main limitations of widespread adoption of VRML has been the require-ment for a VRML plugin. Some time ago these plug-ins were relatively large files (around 3MB), which posed an obstacle for an average user on a slow modem. There were also technical difficulties with viewers for some platforms, including the Macintosh. So if you had VRML content on your web page, there was dan-ger that people would simply go away without seeing your work if they did not have the correct plugin and did not want to take time to install it. Shout 3D avoids the requirement for a plugin. It is composed of platform independent Java applets which are downloaded automatically along with your web3D models. For the first web3D world visited in a particular session the user has to wait for the applet to download (this takes about half a minute on an average modem connection). Subsequent worlds load much faster, because the applet is reused. However, the content one can show with these applets is not as rich as can be offered to users with an installed VRML plugin like Cortona (see *Table 1*). We recommend publish-ing your models using Shout3D, so that it is easier for your potential web audience to check out your work. The more enthusiastic users may then be motivated to install a VRML plugin to better experience your more immersive VRML worlds.

4.3 Java3D

This Application Programming Interface (API) helps Java programmers define three-dimensional objects and put them into immersive interactive virtual 3D worlds. However, some programming time and skill are needed to produce satis-factory results. Note that this approach for programming 3D models directly in Java is different from the Java applets described above, which take VRML models as input and display them.

4.4 Macromedia 3D

Macromedia Director has the capability of handling web3D worlds. One first produces 3D models with some other tool such as Discreet Plasma (see *Table 1*), and then imports them into Director where animation and interactivity are added. At about 3 megabytes, Macromedia's Shockwave3D plugin is almost three times larger than the Cortona VRML plugin, but it provides better texturing and physical world simulations than VRML. Alternatively, it is possible to simulate some 3D effects in Flash. Using Plasma, 3D worlds, including animations, can be exported to Flash as simplified 'drawings' of the 3D models. This approach allows one to produce both animated 2D drawings and web3D models in a short time. Though Plasma is Flash and Shockwave oriented, it also produces excellent VRML output.

4.5 Adobe Atmosphere

This product provides a web3D authoring tool as well as browser for multiuser web3D worlds.

4.6 QuickTime VR

Quicktime is Apple's well known plugin for viewing movies. Since version 3.0, 'QuicktimeVR' can also be used for observing quasi-virtual worlds composed of 'panoramas.' You create a panorama by putting your camera on a tripod and shooting pictures in all directions. For example, you might take about 12 images with the camera positioned horizontally, then you tilt the camera to look for example, 30° downwards to cover bottom objects and so on until you have covered almost everything you can see around you. Then you allow special software (*Table 1*: Sandy Ressler's site) to stitch these pictures together and you obtain a virtual sphere with your own viewpoint at the centre. The walls of the sphere display your pictures, nicely joined together in a seamless panorama which you can rotate or zoom into.

Alternatively, you can also make pictures around some object so that you can observe it from all sides with Quicktime VR. This is more difficult than making the panorama since you need guides to move the camera around the object, while in case of panorama you simply rotate the camera on a tripod.

Quicktime compresses the original panorama very nicely; 10–20 MB of data can be compressed into a into 3–4 MB file. I suggest choosing Cinepak compression from the Quicktime options. But this is still a very large file size for a web project. One way to produce smaller files is to reduce the image resolution—that way you also do not need to shoot as many pictures. Using an extremely wide angle fish eye lens which covers 180°, you take one picture in front of you and one behind to make the entire panorama.

VRML also supports panorama views by letting you put images on the sides of a virtual box containing your model. In some older VRML viewers, panoramas do not move as smoothly as in Quicktime VR. However, the Shout3D panorama viewer in version 2.0 moves almost as smoothly.

5 Conclusion

Virtual reality offers intriguing approaches for visualizing data and communicating ideas. VRML in particular allows us to publish interactive, immersive 3D 'worlds' on the web.

Because VRML is a browser-based technology that allows construction of hyperlinks to other web resources, it is a straightforward process to build a potentially complex VR site by connecting smaller parts. For example, some pages might contain more immersive VR worlds with models and interactivity while the others show more textures, panoramas, etc. Dispersing smaller parts to different web pages makes each page much faster to load and run, and possibly easier to understand. However, this approach can present a problem if it requires your viewers to use too many plugins. Whereas you might use a VRML plugin for virtual worlds, Quicktime for VR panoramas and Flash for nice cartoons, games and fast interactivity, few visitors would be likely to install all three plugins. On the other hand, VRML viewed through Java applets such as Shout3D provide a straightforward way to use a variety of effects without putting undue burden on users.

Reference

1. Pauling, L. (1970). *General chemistry*. W.H. Freeman & Co., San Francisco.

Chapter 11

Web scripting for molecular biologists: An introduction to Perl and XML

Robert M. Horton

Attotron Biosensor Corporation
2533 North Carson Street #A381
Carson City, NV 89706, USA.

1 Introduction

Computer skills are increasingly important to molecular biologists. Internet literacy is essential for finding references in the literature (see Chapter 1), finding gene sequences from the databases (see Chapter 2), and many other tasks (as described in other chapters). Learning how web-based programs work can help you to become a more effective user of these resources.

To organize and analyse your own data, you can often find programs off the shelf (or on the Internet) that meet your needs. However, the nature of research is such that you may encounter situations for which you can't find software that does exactly what you need without some degree of customization.

Developing some basic programming and data management skills can be invaluable if you frequently need to deal with large amounts of computerized data. Most working molecular biologists, especially those using automated large-scale approaches, have probably already reached this conclusion. But beyond the basics of using spreadsheets (1), probably a database management system, and possibly fundamental skills in Unix (2), the choice of where to start in acquiring skills in the bewildering and rapidly developing array of digital technologies can be daunting. This includes deciding which language(s) to learn. Should you learn Visual Basic for Applications so you can add new levels of sophistication to your Excel spreadsheets? Should you learn SQL so you can use databases more effectively? Java? C? C++? C#? MMX vector processing instructions for Intel x86 assembly language? (I'm not entirely kidding about that last one: see (3)).

Though the choices are endless, this chapter presents two computer languages that should probably place high on the priority list of molecular biologists. Perl is a programming language that has many features making it well suited to a variety of Internet- and molecular biology-related tasks. Extensible Markup Language (XML) is a language for structuring data, and the related XSLT is a language for transforming the way such data is structured. Perl and XML complement each other nicely in that Perl can deal with unstructured or loosely structured text data, while XML related technologies provide sophisticated solutions to many common types of problems encountered with structured data.

Each language has a reasonably shallow learning curve, so that knowing even the basics may allow you to apply them to useful tasks. A simple Perl program or XSLT style sheet may be all you need for rearranging data from the format you have it in to the format you need it in. In addition, many Perl scripts and much XML-formatted data are freely available on the Web, so just learning enough to be able to use them can be valuable.

While one chapter cannot really be expected to provide an in-depth tutorial on either language, our aim here is to show how these two languages in particular are of interest to molecular biologists seeking to develop useful computing skills. We will review some background material on each language, and then examine some example programs related to molecular biology and the Internet.

2 Perl

According to its author, the name 'Perl' originated as an acronym for either 'Practical Extraction and Report Language' or 'Pathologically Ecclectic Rubbish Lister' (4, 5). Whether you choose to list rubbish is your own decision, and whether it is pathologic is a judgement call, but Perl's eclectic approach to taking functionalities from other languages helps explain why it is useful for so many applications. It can be used as a 'glue' language to run other programs, so you can use it to perform tasks that might otherwise be done in a Unix shell script, or Windows batch files. Perl is the most popular language for writing server-side web applications; once you know it you can put off learning ASP, JSP, PHP, and a host of other languages that focus on server-side tasks.

As a 'Practical Extraction and Report Language', Perl has a variety of built-in commands and shortcuts that make it a good language to use for dealing with textual data. Most computer-readable data is formatted through the use of keywords, grouping items together on lines or sets of lines, using particular characters (tabs, colons, commas, etc) to separate the fields of a given record, and so forth. Perl is well suited to recognizing patterns in such formatted data, selecting certain data elements ('extraction') and producing output (a 'report') in some other format. Perl subsumes the functions of classic Unix text-file manipulation utilities like sed, awk, and grep—three more little languages you don't need to learn if you know Perl.

Programmers can write Perl modules to interface with C libraries, so using any particular operating system function or set of C-language software tools in Perl is often a matter of finding the correct Perl module. Few molecular biologists should need to program in C themselves.

Perl supports object-oriented programming, but unlike a more pure object-oriented language such as Java, objects are optional in Perl. If you have a background in some other procedural programming language, learning Perl will be a matter of learning syntax and tricks, but it won't require you to learn a whole new paradigm. Perl programs are typically much more concise than equivalent programs written in many other languages. This is especially true for tasks for which Perl has good built-in support, like dealing with character-based data. Perl's conciseness can be a blessing when it helps you accomplish tasks more quickly, but it can also be a curse when carried to extremes, because it can be daunting to try to figure out what someone else's Perl program is doing. In Perl, it is both easy and tempting to write 'obfuscated code' that will be difficult for other programmers to figure out. Two things mitigate this curse: first, no language is immune to tangled code. Second, obfuscation is a much greater problem in larger programs. Starting with small scripts lets you develop organizational programming habits, including use of comments and built-in documentation, descriptive variable names, reusable functions and subroutines, and objects when appropriate. These organizational skills will serve you well when you start to build larger programs, even in other languages.

Perl can be used for client-side scripting inside a web page, but this is rarely done because it requires a browser plug-in (available from the ActiveState website, see Web Resources). Javascript (see Chapter 9) is generally the language of choice for scripting a web browser, and Java is currently the dominant language for more complex programs that need to run inside a web page. While some simple molecular biology tools can be completely browser-based (6), in general, client-side programming is more useful for user interface issues rather than more computational or data-centered tasks. Most sophisticated web applications have both client and server components. Perl is extremely popular for server-side web programming.

Perl is generally used with a language interpreter, rather than a compiler. This means you don't have to go through a separate compilation step every time you make a change to your program (as you do with Java, C, C++, etc.). Interpreted languages thus have an advantage for those of us who (still) rely to some extent on trial and error. While other interpreted languages (most notably Python) have a following in bioinformatics, the widespread popularity of Perl is a selling point because of the extensive community resources devoted to the language. These include numerous web-based tutorials and references (see Web References), as well as published books targeted towards molecular biologists (7).

We now proceed to our first example, which shows a simple application of 'extracting' and 'reporting' data.

2.1 Example 1: Extracting atomic coordinates from a PDB file to use in a VRML model

Inspired by Chapter 10, we'll use simple scripts to help us build a VRML model of a protein. To this end, we want to extract the *x*-, *y*- and *z*-coordinates of all of the alpha carbons from a crystal structure represented in a Protein Data Bank (PDB) file. An extract from a PDB file is shown in *Figure 1(a)*. The beginning of a PDB file contains information such as journal references for the authors; here, we are primarily interested in the atomic coordinates as shown in the lower portion of the window. In the third column of this part of the file, the letters 'CA' denote the alpha carbons, while the seventh, eighth, and ninth columns contain the *x*-, *y*-, and *z*-coordinates of the atoms.

Figure 1(b) shows a very short Perl script used to extract just the lines describing the alpha carbons of the protein backbones, and to format them so they can be further manipulated in a spreadsheet. This simple example illustrates many concepts of Perl programming. The first line starts with the pound sign, '#'. This symbol is used to mark comments in Perl; anything following a pound sign, all the way to the end of that line, is ignored by the program. Line 1, where the pound sign is followed by an exclamation point, is a special case; it is ignored by the program *per se*, but it gives information to the operating system about how to run the interpreter that runs the program. It specifies exactly which Perl interpreter to run, and which options to run it with. On personal computers, it is generally not necessary to specify which Perl interpreter to run, because you probably only have one. The option '-w' turns on 'warnings', and the following line turns on a 'strict' set of programming rules. These cause the Perl interpreter to notify us if it notices any of a set of potential errors in the program. It is generally a good idea to run Perl with warnings and the strict rules turned on, at least during development. These give you feedback about potential errors you may have in your program, such as using an undeclared variable. These warnings can be very useful; for example, you may have misspelled the name of a variable, so that it appears as a new, undeclared variable. Without the warning, Perl would just treat it as a new variable and not complain, so your spelling error might escape immediate notice.

This program is written as a 'filter', which is a program that takes its input from 'standard input', and sends its output to 'standard output'. When running from the command line in Windows or Unix (including Macintosh OS X), the standard input is the keyboard, and the standard output is the screen. In both Unix and Windows, a file can be directed to the standard input of a program using the input redirection operator, '<', at the command line. Similarly, output can be redirected to a file using one of the output redirection operators, '>' (to write a new file) or '≫' (to write, or append if the file already exists). Writing programs as filters that use the standard input and output has two advantages. First, it makes our programs simpler because we do not have to worry about

(a)
(b)
(c)
(d)

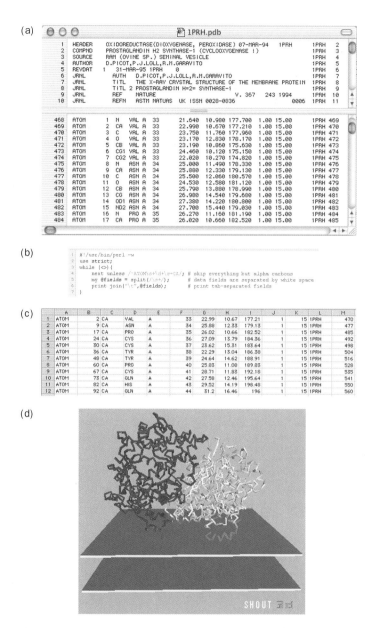

Figure 1 (a) Selected portions of an atomic coordinate file in Protein Data Bank Format. This image was captured of a text editor that lets you split the window, so one portion of the file (lines 1–10) is shown in the top part, and a different portion of the same file (lines 468–484) is shown in the bottom part. (b) A Perl script to extract the coordinates of the alpha carbons from a PDB file and format them for pasting into a spreadsheet. (c) Output of the program after pasting into a spreadsheet. Only the first few lines of output are shown; the entire output contained 1108 lines. (d) A 3D model of a protein and its relationship to the cell membrane. Here each chain of the protein is represented as a VRML extrusion node as described in Chapter 10. Planes have been added to represent the surfaces of a cell membrane.

opening and closing files. Second, a series of filters can be chained together in a 'pipe' from the command line, where the output of one filter is sent or 'piped' to the input of another. While that topic is beyond the scope of this chapter, it is covered in most introductory books on Unix. The script from *Figure 1(b)* was saved in a file named 'CAextractor.pl'., the input file was '1PRH.pdb' from the PDB database, and the results were saved to a new file called 'pghs.out'. The command line to run the program was:

```
perl CAextractor.pl < 1PRH.pdb > pghs.out
```

The program has selected the data for only the alpha carbons. The tabs inserted in the blank spaces cause the fields within each line to be placed into columns when they are pasted into a spreadsheet, as shown in *Figure 1(c)*. Columns G, H, and I contain the *x*-, *y*- and *z*-coordinates. VRML code is constructed by cutting and pasting these columns into a file containing the framework of an Extrusion node, as described in Chapter 10. A separate Extrusion is used for each peptide chain. The coordinates indicate 'spine points' through which the extrusion is traced, resulting in a simple 3D model of the protein backbone.

The grand result is shown in *Figure 1(d)*, where the traces of the backbones of each chain in a protein are used to model the relationship of the protein to the surface of the cell membrane (8, see also 2D illustration in Web References). Semi-transparent planes have been added to represent the cell membrane surfaces; you can see portions of the protein sticking through them. This model is intended primarily to illustrate the idea that VRML (unlike the dedicated molecular viewing programs described in Chapter 9) can be used to mix molecular and non-molecular structures. Please don't take it too seriously as saying anything about cyclooxygenase!

With a very slight modification, we could have our little program print just the columns of interest. Instead of printing the whole '@fields' array for each alpha carbon, we can specify which items we want by using an 'array slice'. Note that, as in many computer languages, the elements in an array are numbered starting with zero. Thus, the seventh, eighth, and ninth elements are given by '@fields[6..8]'. Changing line 6 in the program of *Figure 2(b)* to

```
print join("\t",@fields[6..8]);
```

will output just these elements.

This simple program shows the basis of 'extraction and reporting'.

2.2 Server-side programming on the web

In a simple view of the Internet, a computer can act as a 'client' (usually the computer asking for information) or a 'server' (a computer supplying answers to client's questions). These terms also refer to programs; that is, a client program is one asking for information, and a server program is one supplying answers. A web server is an example of a server program. In general, we will use these terms to mean the programs rather than the computers, because you can have more than one program running on a computer. For example, you might have both a web browser and a web server running on your laptop, and you can use your browser to ask your web server for information even if your laptop is not connected to the Internet. Similarly, programs on web pages can run 'client-side', in which case the CPU on your PC is doing the work, or 'server-side', on whatever computer is running the web server (usually, but not necessarily, somewhere out on the Net).

Traditionally, most server-side web programming has been done using the Common Gateway Interface (CGI), a standard way of communicating between a web browser and another program (usually called the 'CGI program') running on the same server. The web server recognizes requests that need to be run by the CGI program; the web server calls the CGI program, passes it the request using the standard protocols of the CGI, receives the results, and sends them back to the client. The CGI approach is very flexible because programs can be written in almost any programming language. They run separately from the web server, and only need to know how to communicate with the server in order to receive requests and send back replies.

Microsoft Active Server Pages (ASP), Java Server Pages (JSP), and the PHP Hypertext Preprocessor (now officially not an acronym), are template directed approaches to server-side programming. Instead of a program that writes a web page, this approach is more like a web page that has some programming in it. The use of templates by these languages is in some ways similar to XSLT, which is described later in this chapter. These environments generally run inside the web server instead of as external programs communicating through CGI. JSP, of course, uses Java. PHP is a language in its own right, with a syntax similar to a cross between Perl and C. ASP can use several languages, including Visual Basic, JScript (Microsoft's flavor of Javascript), and Perlscript (a Perl interpreter that can be added to a scripting host such as ASP). We will not consider these technologies here other than to point out that learning Perl will help you to use ASP, and to a degree PHP, if the need arises.

One criticism of CGI is that launching the CGI program, passing it the request, and getting the results back, all take time over and above that required for actually processing the request. In some circumstances this overhead may take more time than the processing. Fortunately for Perl programmers, a module has been developed that adds a Perl interpreter to a web server. This allows most Perl programs originally written as CGIs to run inside the web server with little or no change to the Perl program code. Switching to the use of this mod_perl web server module rather than an external CGI program is essentially a matter of web server configuration (9).

Here we present an example of a simple Perl CGI program.

2.3 Example 2: A Perl CGI program to generate random protein or DNA sequences

The code and outputs of this example are shown in *Figure 2(a–d)*. Before wading through the detailed description of the code, you may want to look at panels c and d of the figure to see the results we are trying to achieve.

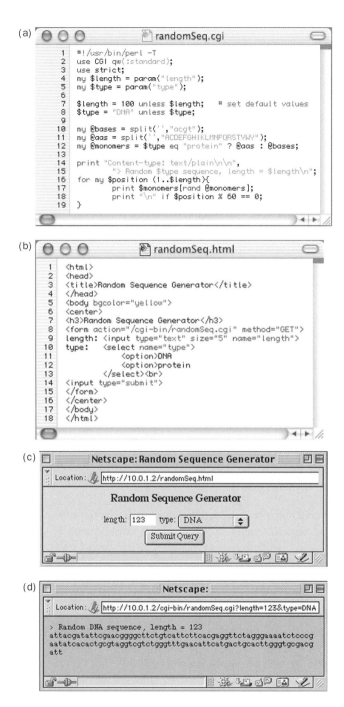

Figure 2 (a) Perl CGI script 'randomSeq.cgi'. (b) HTML for a web page that serves as an interface to the Perl CGI program. (c) The web page as seen in a browser. (d) A random sequence generated by the CGI program.

Figure 2(a) gives the Perl code for the CGI program. Please take a moment to look over the program briefly; I hope the fact that it is a mere 19 lines long will provide some solace as we dive into several pages of explanation.

Line 1 specifies exactly which Perl interpreter we want to use; the default location on Mac OS X is '/usr/bin/perl'. In many cases, this is the only line of a script you may need to change in order to run it on a different computer. This line also lets you specify 'flags' that control special features of the Perl interpreter. Here we have included the '-T' flag, which turns on 'taint checking'. Tainting is a security feature that helps to prevent user-entered data from being used in a dangerous way. For example, you probably don't want to let all visitors to your website run any command they choose on your computer. Data that originated from a website user will be considered 'tainted', and will have certain restrictions placed on the ways it can be used. Note that the '-w' flag, which we used in *Example 1* to turn on warning messages, is not used here. In general, it is a good idea to turn warnings on, but we do not in this case because in a CGI program warning messages just get sent to the web server's error log. Line 2 invokes the 'strict pragma', which enforces a number of rules that are useful for preventing a variety of programming errors. Line 3 imports Perl's 'CGI' module, which defines a set of functions that make it much easier to write CGI programs. We import the 'standard' set of functions in this example; in fact, the only function we use from the CGI module is called 'param()'. Note that the key word 'my' is used when a new variable is first declared. Many Perl scripts do not use this kind of declaration, but it results in a warning if 'strict' rules are in effect. In lines 4 and 5, this function retrieves the named parameters that the user sends to the program (we will see below how these parameters are sent). The two parameters named 'length' and 'type' are copied into similarly named scalar variables. A scalar variable holds a single value, and its name begins with a dollar sign. Lines 7 and 8 assign default values to these variables if no values were sent by the user. Line 10 establishes an array of the bases that comprise DNA sequences. The bases are listed in a string, which is 'split' into an array using an empty string as the pattern to split on. A similar array is created for the amino acids in line 11. Which array we use to build our random sequences is determined in line 12: the '@monomers' array will be set to the amino acids if the $type variable is 'protein', and to the DNA bases otherwise.

The CGI protocol gives you the option of generating a variety of different types of outputs. For example, a CGI script might create a web page, or draw an image representing a chart of some type of data. In line 14 we declare that the type of content we will produce will be plain text. This first line of our program's output is really the last line of the hypertext transfer protocol (HTTP) header. The header is not seen by the user, only by the user's web browser program. The header must be followed by a blank line so the browser will know the header has ended and the page content is about to start. We add this blank line by adding an extra newline character at the end of the content-type line. This newline symbol is represented by '\n'; the backslash is an 'escape' character which means the character following it has a special meaning—it is no ordinary 'n'. Line 15 gives another line of output for the print command to use. Note that Perl's print command takes a comma-separated list of arguments, so we can print multiple things at once. This second output is actually the first part that will be visible in the browser. Since we are putting the random sequence in FASTA format, we use the first line, which starts with a greater-than sign, to make a comment about the sequence. Note that the variables $type and $length are written right into the comment line. These variables will be 'interpolated', that is, their values will be substituted in place before the line is sent to the output. (Full quotation marks are often called 'double quotes' to distinguish them from 'single quotes'. In Perl, strings in double quotes have their variables interpolated, while single quoted strings do not.)

Lines 16 through 19 do the business of constructing the random sequence. The loop goes through every sequence position from the beginning to the specified length and chooses a random monomer to use at that position. Note that an individual monomer in the @monomers array is itself a scalar (single-valued) variable. Thus, the monomer at position number 2 in the array is referred to as $monomers[2]. Elements of arrays in Perl (as in C, Java, and many other languages) are numbered starting at zero, so $monomers[2] is 'g' if the sequence is DNA. The randomness is introduced by Perl's 'rand' function, which takes a number as an argument and returns a pseudorandom number between zero and the given number. But note that we have supplied the 'rand' function not with a number, but with an array as a parameter. Perl's arrays are smart enough to know that when they are used in a 'scalar context', that is, where a scalar value like a number is expected, they produce a number representing the length of the array. This may seem confusing at first, but it is very concise. Another subtle issue is that the rand function produces real numbers, not integers. When a real number is used as an index of an array, it is truncated to the whole number part (1 in this case). Thus, $monomers[1.79926509782672] is 'c'. Finally, line 18 adds a bit of formatting by adding new line characters to the output of every 60 characters. This is done using the modulus operator (the percent sign).

There is not really very much that is special about this program that distinguishes a CGI program from an ordinary Perl script. We use the param() function of the CGI module to get our input parameters, and we had to be sure to write the content-type line (followed by a blank line!) as the first line of our output, but that is really about all that is different from a simple script. In particular, note that we did not say where to print the output; we send everything to the default (standard) output, and the web server sends it back to the user.

Now that we have seen the details of the server side of the program, we'll look at how the user specifies the parameters for the CGI program to use. Since CGI is a web-based technology, we construct the user interface in a web page. The web page as viewed in a browser is shown in *Figure 2(c)*, while the HTML code to generate that web page is shown in *Figure 2(b)*. It consists of run-of-the mill HTML (see Chapter 9) with the addition of a 'form' (lines 8 through 15). Line 8 holds the tag that begins the HTML form. The 'action' attribute of the form tag specifies the URL where the data will be sent once the form is filled out. In this case, it will go to our CGI program, which we have put on the same web server as the form itself. Since the form and the CGI program are on the same server, the URL is relative, rather than absolute, and does not need to contain the `http://` and host name parts of a full (absolute) URL. The browser can reconstruct these parts when it needs to, since they are the same as for the web page that is sending the data. Using a relative URL makes it easier to move the form and CGI program to a different server. The second attribute is the method by which the form data will be sent to the server. The two common methods are 'GET' and 'POST'. In the GET method, the parameter names and values from the form are appended to the end of the CGI program's URL. This can be seen in the location box of the result page shown in *Figure 2(d)*: the URL of the CGI program has a question mark and two parameter values (`length=123&type=DNA`) attached.

If you change the form method from 'GET' to 'POST', the form information will no longer be sent attached to the end of the URL; it goes behind the scenes. The POST method has the advantage that it can be used to transmit larger amounts of information (there are only so many characters of data you can piggyback onto a URL). However, the GET method has the advantage that all the information submitted by the form is encapsulated right in the URL. This lets you do things like save a search as a bookmark, and repeat the search every time you visit the site (these are the 'smart links' described in Chapter 7). You can also change the parameter values right in the location box and reload the page. In this way, you can run the CGI program without even using the HTML form. This approach makes it easy for other programs to send requests to a CGI program, since it is mostly a matter of building the right URL string.

It is a valuable educational experience to set up a simple CGI script like this one and a form to interface with it. Most web servers place some restrictions on CGI scripts for security reasons. For example, servers are commonly configured to only allow CGI programs to be run in certain directories (where 'cgi-bin' maps to, for example). To set up a CGI program of your own, you will need to consult the documentation for your web server, or seek help from your system administrator. To generate *Figures 2(c) and 2(d)*, the CGI and the interface form were loaded on a web server running on a machine on a Local Area Network (LAN). The server on the LAN can be identified by an 'IP number' (like 10.0.1.2), rather than a name. This particular IP number is only usable from within the LAN.

Once your CGI script is running, you will probably want to modify it to change its behaviour or to add new features. One interesting modification to this example might be to change the strings that specify which bases and amino acids are allowed. You could produce random RNA sequences, for example. More interestingly, changing these strings lets you change the relative frequencies of the various monomers. For example, the string 'acgt' has equal numbers of each base, so random sequences generated from the resulting monomers array will contain, on average, roughly equal frequencies of each base. Changing this string to 'aacgt' (or 'acgta', etc.) doubles the relative probability of picking 'a' at any given position. Or, changing the string to an actual gene sequence would change the probabilities to reflect the base composition of that sequence. With proteins, it might be useful to use the 'average' composition of a protein sequence, which can be found on the ExPASy server (see Web References). This average composition is given in terms of percentages for each amino acid, to the nearest tenth of a percent (Alanine 8.3%, Arginine 5.7%, etc.) . Thus, one could reflect this data in a 1000 amino acid-long sequence with 83 'A's, 57 'R's, and so on.

Protocol 1
Running Perl on Windows, Unix, or Macintosh

The examples in this chapter were tested with ActivePerl version 5.6 on Windows ME, MacPerl 5.2.0r4 on MacOS System 8.5, and Perl version 5.6.0 on Mac OS X.

Windows

1   Download Perl from the ActiveState website (`www.activestate.com`) and run the installer.

2   Enter a Perl script in a text editor, and save it as text (i.e. not RTF, not Word format; just plain old ASCII text).

3   Open a command line window (this was called the 'DOS prompt' on Windows ME and earlier) and run the script with a command such as

```
perl myScript.pl < myInput.txt > myOutput.txt.
```

The input/output redirection and piping operators ('>','<', and '|') are described in the text; they work similarly on Windows, Unix, and Mac OS X.

Unix or Macintosh OS X

1   Install Perl. For Mac OS X, Perl is on the Apple Developer CD-ROM. For Linux, it can be installed from the distribution CD-ROMs when the operating system is installed, or by using your distribution's package manager. If you are willing to compile it yourself, the latest source code can be found at `www.perl.com`. On multiuser Unix systems, contact your administrator if Perl is not already installed.

2   Enter a Perl script in a text editor, and save it as text. Be sure to use Unix line breaks (see protocol 3); if you use Macintosh line breaks, nothing happens when you try to run the script!

3   (optional) change the permissions on your file to make it executable, using the command `chmod 755 myScript.pl` (this assumes, of course, that your script is named 'myScript.pl'). That way, you don't need to type the word 'perl' in front of your script to run it.

4 Open a terminal window and run the script with a command such as

```
perl myScript.pl < myInput.txt > myOutput.txt
```

The Terminal application in Mac OS X is found under 'Applications:Utilities'.

On operating systems designed to run with multiuser users, like Unix and Linux, there might be several different versions of Perl interpreters installed, and you need to be very specific about which one you mean. The example in *Figure 2* assumes the Perl interpreter is named 'perl', and it resides in the directory '/usr/bin', a common place to put Perl.

If you run this script on a Unix machine (including OS X), you need to be sure this first line actually points to where Perl is on that machine. To find out, type the command 'which perl'. It will respond with something like '/bin/perl'. Be sure to change the first line of the script to match this location! Note also that the scripts in this chapter use Perl version 5. To find out the version of the interpreter you are specifying, type its name with the '-v' option. For example, if 'which perl' returned '/bin/perl', we'd type '/bin/perl-v' to find out which version this is. If it is something less than 5, please ask your system administrator to help you locate a version 5 interpreter.

Macintosh (OS 8 or 9)

1 Download and install MacPerl (www.macperl.com)

2 Launch the MacPerl application.

3 Open, enter, or paste a Perl script in the MacPerl editor (choose 'new' or 'open...' under the 'file' menu).

4 Run the script by selecting the second 'run' command from the 'script' menu (i.e. 'run "filename"').

5 To run a script as a 'filter' that reads data from standard input, save it as a 'Droplet'. Drag and drop text files on the droplet for processing.

Since there is no command line on the traditional Macintosh OS, a MacPerl script can be saved as a 'droplet'; a text file dropped on the program's icon will be read from standard input, and the output is sent to a console window.

Protocol 2

Transferring files in ASCII mode

A common cause of frustration for beginning Perl programmers is actually not even Perl's fault; it stems from the fact that different operating systems use different characters to

Protocol 2 continued

mark the ends of lines in files: the Macintosh Classic OS uses '^M' (control-M, a carriage return, ASCII character 13), Unix and Mac OS X use '^J' (linefeed, ASCII character 10), and Windows uses ^M^J (both). For more gory details see (5).

In general, Perl handles these differences behind the scenes: MacPerl will split the contents of a file into lines using the Mac character, etc., and the command 'print "\n"' will mark the end of the line with the character(s) appropriate to the system you are using. This is the good news; it means you can run many Perl programs written for Unix on a Mac or Windows machine without changing anything.

Problems can occur if you somehow transfer a file between different types of computers without translating the line breaks. This can happen, for example, if you use FTP in binary mode instead of ASCII mode to upload a data file from your Windows PC to a Unix web server; ASCII mode translates the line breaks, but binary mode does not. It can also happen if you save a text file from your Macintosh to a DOS formatted disk, and copy it to a Windows PC.

If you try to run a Perl script that has its line breaks marked with the incorrect characters, it will usually fail, and may give mysterious error messages. If your script works properly, but your data file has the wrong line breaks, you may get weird results. For example, your program may end up treating the whole file as one line when you were expecting it to treat it as a collection of lines. The program in *Figure 2*, for example, would find no lines beginning with 'ATOM', and would produce no output.

This problem can be insidious because the offending linefeed and carriage return characters may be invisible in some common text editors. The best defence is to use a good programmer's text editor that can show you invisible characters; some such programs are indicated in the Web Resources Table. This also makes it easier to tell 'soft-wrapped' lines from 'hard-wrapped' lines at a glance.

Finally, we come to a brief protocol for ASCII mode file transfer using FTP:

1 Open an FTP connection to the server where the file resides. On Windows and Macintosh machines, graphical FTP programs (e.g. WS_FTP and Fetch, respectively) let you choose servers and options using menus and forms. For old-fashioned command line ftp (which ships with Unix, Mac OS X, and Windows), enter the command 'ftp' followed by the Internet address of the server.

2 Enter your username and password as requested. Some servers allow 'anonymous FTP', usually by entering the username 'guest', with your own email address as password.

3 Enter ASCII mode; in graphical FTP programs, there should be a checkbox to specify 'ASCII' (or 'text') versus 'binary' mode. For command-line programs, enter 'ascii' at the ftp prompt to turn on ASCII transfer mode (enter a question mark to see a list of FTP commands).

4 Select the data file to transfer. Graphical programs usually use either drag and drop, or let you select files from the list on the remote host, then click an arrow to bring them to your local host. The old-fashioned command is 'get', followed by the desired file name. The command 'mget' retrieves multiple files.

Protocol 3

Perl 'one-liner' to steal JavaScript

1 Type the following at the command prompt (all on one line):

```
perl -MLWP::Simple -e "print get 'http://www3.ncbi.nlm.nih.gov/
entrez/query/query.js'"
```

2 Press the return key. The contents of the file will be displayed.

This is a Perl script specified entirely on the command line (a 'one-liner'). It fetches a file from the given URL. This can be useful for 'hidden' files, like some Javascript source files. Since these files are not directly linked from other pages, you can't right click on their link to save them as with normal web pages. The '-M' flag loads a Perl module, in this case, the 'LWP' module (Library for Web access in Perl). This module handles the hard part; you just tell it to get a URL.

3 Extensible Markup Language (XML)

3.1 Background

Ordinary conventions of spelling notwithstanding, the 'X' in 'XML' stands for 'eXtensible'. Unlike HTML, XML can be 'eXtended' in that new tags can be specified for particular uses. A collection of such specialized tags can comprise a specialized markup language in its own right. For example, the X3D 'modeling language' comprises a specialized set of XML tags that can be used to specify 3D structures, similar to what one might do with VRML (see Chapter 11). An X3D representation of a 3D model, however, is an XML file. Superficially, the difference between a VRML model and the corresponding X3D model is a matter of syntax. But being a form of XML means that X3D models can be dealt with using XML tools. The fact that so many kinds of data can be structured using XML means that XML tools in general are widely useful. This encourages development of tools with wide applicability; you may use XML tools to transform an X3D model, for example, even if the developers of that tool had no idea it would be used for 3D models.

XML is a 'meta-markup language' which can be used to specify more specialized languages. Some such specialized XML markup languages of interest to scientists include Scalable Vector Graphics (SVG), an XML-based language for specifying sophisticated 2D graphics, Chemical Markup Language (ChemML), MathML for mathematics, and MicroArray Gene Expression Markup Language (MAGE-ML); see Web References for more details.

Fundamentally, XML is used for specifying how the structure of (potentially complex) data can be recorded along with the data, in a (more or less human-readable) text file. XML provides a way to include 'metadata' along with the data in a text file. 'Metadata' is some type of data about the data, such as how the elements of data relate to one another. The metadata is represented by 'markup tags' that identify the type of data represented by a particular element. An example is given in *Figure 3(a)*, which will be described in detail below. These tags will appear familiar to people who know the HTML used to create web pages (see Chapter 9).

XML can be applied to essentially any type of hierarchically structured data. In computer science, a hierarchical data structure in which any data element 'belongs' to a higher-order element, with the exception of the highest-order element of all, is called a 'tree', and the highest-order element is called a 'root'. (This terminology assumes that the 'tree' is upside-down, with the root at the top.)

(a)

```
     XMLlit.xml                                            _ □ ×
 1  <?xml version="1.0"?>
 2  <PubmedArticleSet>
 3  <PubmedArticle>
 4     <MedlineCitation Status="In-Process">
 5        <MedlineID>21883721</MedlineID>
 6        <PMID>11886702</PMID>
 7        <DateCreated>
 8           <Year>2002</Year>
 9           <Month>03</Month>
10           <Day>11</Day>
11        </DateCreated>
12        <Article>
13           <Journal>
14              <ISSN>0169-2607</ISSN>
15              <JournalIssue>
16                 <Volume>68</Volume>
17                 <Issue>1</Issue>
18                 <PubDate>
19                    <Year>2002</Year>
20                    <Month>Apr</Month>
21                 </PubDate>
22              </JournalIssue>
23           </Journal>
24           <ArticleTitle>Complexity of biomedical da
25           <Pagination>
26              <MedlinePgn>49-61</MedlinePgn>
27           </Pagination>
28           <Abstract>
29              <AbstractText>Atrial fibrillation (AF)
30           </Abstract>
31           <Affiliation>Department of Medical Inform
32           <AuthorList>
33              <Author>
34                 <LastName>Dugas</LastName>
35                 <ForeName>M</ForeName>
36                 <Initials>M</Initials>
37              </Author>
38              <Author>
39                 <LastName>Hoffmann</LastName>
40                 <ForeName>E</ForeName>
41                 <Initials>E</Initials>
```

Figure 3 (*Continued*)

(b)

```
biblio_htm.xsl

1  <?xml version="1.0"?>
2  <xsl:stylesheet version="1.0" xmlns:xsl="http://www.w3.org/1999/XSL/Transform">
3  <xsl:output method="html" indent="yes"/>
4    <xsl:template match="/PubmedArticleSet">
5      <HTML>
6        <HEAD><TITLE>PubMed results</TITLE></HEAD>
7        <BODY BGCOLOR="#9999FF">
8          <H1 ALIGN="CENTER">PubMed results</H1>
9          <OL>
10           <xsl:apply-templates select="PubmedArticle/MedlineCitation"/>
11          </OL>
12        </BODY>
13      </HTML>
14    </xsl:template>
15    <xsl:template match="MedlineCitation">
16      <LI>
17        <xsl:apply-templates select="Article/AuthorList"/><!-- authors -->
18        <B><xsl:value-of select="Article/ArticleTitle"/></B>
19        <xsl:text> </xsl:text>
20        <I><xsl:value-of select="MedlineJournalInfo/MedlineTA"/></I>
21        <xsl:text> </xsl:text>
22        <B><xsl:value-of select="Article/Journal/JournalIssue/Volume"/></B>:
23        <xsl:value-of select="Article/Pagination/MedlinePgn"/>,
24        <xsl:value-of select="Article/Journal/JournalIssue/PubDate/Year"/>.
25        <A>
26          <xsl:attribute name="HREF">
27            <xsl:text>http://www.ncbi.nlm.nih.gov:80/entrez/query.fcgi?</xsl:text>
28            <xsl:text>cmd=Retrieve&amp;db=PubMed&amp;dopt=Abstract&amp;</xsl:text>
29            <xsl:text>list_uids=</xsl:text>
30            <xsl:value-of select="PMID"/>
31          </xsl:attribute>
32          <xsl:text>Pubmed</xsl:text>
33        </A>
34      </LI>
35    </xsl:template>
36    <xsl:template match="Article/AuthorList">
37      <xsl:for-each select="Author">
38        <xsl:value-of select="LastName"/>
39        <xsl:text> </xsl:text>
40        <xsl:value-of select="Initials"/>
41        <xsl:if test="position() != last()">, </xsl:if>
42        <xsl:if test="position() = last()">. </xsl:if>
43      </xsl:for-each>
44    </xsl:template>
45    <xsl:template match="*"></xsl:template><!-- default: do nothing -->
46  </xsl:stylesheet>
```

(c)

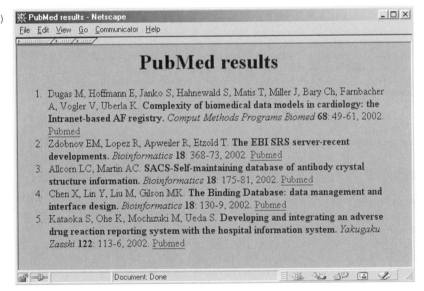

Figure 3 (*Continued*)

(d)

biblio_rtf.xsl `_ | □ | x`

```
1  <?xml version="1.0"?>
2  <xsl:stylesheet version="1.0" xmlns:xsl="http://www.w3.org/1999/XSL/Transform">
3  <xsl:output method="text" encoding="UTF-8" indent="no" />
4
5  <xsl:template match="*"><!-- default: do nothing --></xsl:template>
6
7  <xsl:template match="/PubmedArticleSet">{\rtf1 {\fonttbl {\f0 Times New Roman; }}
8  \fs28 \pard {\b PubMed results:\par } \fs24
9  <xsl:apply-templates/>
10 }
11 </xsl:template>
12
13 <xsl:template match="PubmedArticleSet/PubmedArticle">
14 \pard {\par <xsl:value-of select="position() div 2"/>. }<xsl:apply-templates/>
15 </xsl:template>
16
17 <xsl:template match="MedlineCitation">
18 <xsl:apply-templates select="Article/AuthorList"/>
19 {\b <xsl:value-of select="Article/ArticleTitle"/>}
20 {\i <xsl:value-of select="MedlineJournalInfo/MedlineTA"/>}
21 {\b <xsl:value-of select="Article/Journal/JournalIssue/Volume"/>}{:}
22 { <xsl:value-of select="Article/Pagination/MedlinePgn"/>, }
23 {<xsl:value-of select="Article/Journal/JournalIssue/PubDate/Year"/>}
24 { (PMID <xsl:value-of select="PMID"/>)}
25 </xsl:template>
26
27 <xsl:template match="AuthorList">
28 <xsl:for-each select="Author">
29 {<xsl:value-of select="LastName"/>, <xsl:value-of select="Initials"/>}
30 <xsl:choose>
31     <xsl:when test="position() = last()">{. }</xsl:when>
32     <xsl:when test="position() = last() - 1">{, and }</xsl:when>
33     <xsl:otherwise>{, }</xsl:otherwise>
34 </xsl:choose>
35 </xsl:for-each>
36 </xsl:template>
37
38 </xsl:stylesheet>
```

(e)

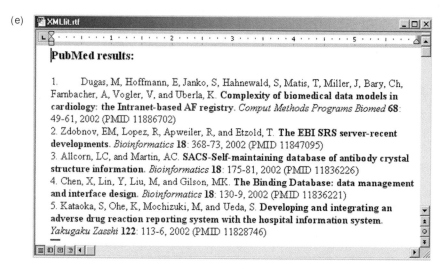

XMLlit.rtf `_ | □ | x`

PubMed results:

1. Dugas, M, Hoffmann, E, Janko, S, Hahnewald, S, Matis, T, Miller, J, Bary, Ch, Farnbacher, A, Vogler, V, and Uberla, K. **Complexity of biomedical data models in cardiology: the Intranet-based AF registry.** *Comput Methods Programs Biomed* **68**: 49-61, 2002 (PMID 11886702)

2. Zdobnov, EM, Lopez, R, Apweiler, R, and Etzold, T. **The EBI SRS server-recent developments.** *Bioinformatics* **18**: 368-73, 2002 (PMID 11847095)

3. Allcorn, LC, and Martin, AC. **SACS-Self-maintaining database of antibody crystal structure information.** *Bioinformatics* **18**: 175-81, 2002 (PMID 11836226)

4. Chen, X, Lin, Y, Liu, M, and Gilson, MK. **The Binding Database: data management and interface design.** *Bioinformatics* **18**: 130-9, 2002 (PMID 11836221)

5. Kataoka, S, Ohe, K, Mochizuki, M, and Ueda, S. **Developing and integrating an adverse drug reaction reporting system with the hospital information system.** *Yakugaku Zasshi* **122**: 113-6, 2002 (PMID 11828746)

Figure 3 (a) Pubmed results in XML format. Indentation and line breaks have been added to make the structure more clear. Only the first lines of the large data file are shown here. (b) An XSL stylesheet to format PubMed results as a web page. (c) Screen shot of output from applying the stylesheet from part (b) to the data of part (a), viewed in a web browser. (d) XSL stylesheet to format Pubmed results as a Rich Text Format (RTF) document. (e) Screen shot of output in RTF, viewed in Microsoft Word. This file was generated by applying the stylesheet in (d) to the data in (a).

Thus, a phylogenetic tree in biology can be described as having a tree structure. Each element is called a 'node', and a node belonging to a higher-order node is called a 'child' node.

Tree structures can be used to describe an amazing variety of things. Any computer program, for example, can be structured into a 'parse tree' that specifies how the statements relate to one another. Tree structures can describe things from algebraic equations to sentences in human languages. They are used to represent many kinds of documents, from web pages to Microsoft Word documents. The fact that 3D models can be described as trees explains why the different types of shapes and features used in building 3D worlds are called 'nodes' (see Chapter 11).

Consider the structure of a typical scientific paper as an example. There is a single root for the whole document, which has a single title, a list of authors, an abstract, the body of the article, and a list of references. The list of authors can be represented as a node, with a child node for each author. An author node has both the author's name, and an affiliation. The body of the article can be a node, and its child nodes would contain an introduction, a materials and methods section, experimental results, discussion, etc. There are a variety of ways you could structure the document (e.g. you might leave out the body node and make the introduction, etc., child nodes of the document root), but they are still essentially tree structures.

Tree structures are very flexible in that they can also be used for data that is not really very structured. For example, a collection of elements that do not really contain any other elements can be represented as a short, one-layer tree with a root node (the collection) and a bunch of child nodes (all the elements). Some structures cannot be easily represented in trees, however. If a descendent of a node is also the parent of that node, or if a node is a child of more than one parent, the structure is a 'cyclic graph' rather than a tree. But in general, trees can represent a huge variety of interesting data structures, and XML can represent all kinds of tree structured data. Some structures more complicated than trees can be represented in XML using various tricks that are beyond the scope of this chapter.

Once the structure of a data set has been marked up in XML, we do not need to use the kinds of Perl pattern-matching tricks described in the first part of this chapter to extract particular elements of the data. This is because a program can just read the tags, and use the hierarchical relationships between tags, to identify particular elements of interest. There are Perl modules that simplify the task of reading an XML file and structuring the data in the specified tree relationship so your program can access it. However, we will focus on a different approach, using a language called XSLT to manipulate XML data.

The Extensible Stylesheet Language (XSL), allows you to write a 'stylesheet' that specifies how a particular XML document should be presented. If you have used stylesheets in a word processor to control the appearance of a document, or Cascading Style Sheets (CSS) with a web page, you may have some familiarity with the general concept. Stylesheets let you separate the content of a document (what it says) from the presentation (how it appears on the page, fonts, text size, text colour, etc.).

Stylesheets in XML can be considerably more powerful than just controlling formatting. In addition to formatting, they can control 'transformation' of a structured XML document into another structure. For example, a style sheet might act as a filter to only select parts of the input document for inclusion in an output document. Or it might be used to change the order of elements in a document, such as by sorting them. To distinguish these types of transforming stylesheets from ordinary formatting stylesheets, the language used to write these stylesheets is called 'XSLT', or XML Stylesheet Language for Transformations.

XSLT stylesheets use a template-driven, rule-based architecture. It is a 'declarative' language in that you 'declare' what output should be produced when a certain pattern is encountered. Most programming languages, in contrast, are 'imperative', in that the programmer must explicitly tell the computer what to do in a sequence of steps.

XSLT is an example of an XML-based language that can be used for programming. Thus, XSLT stylesheets are actually programs that operate on XML documents, and they are themselves XML documents.

The XSLT language is specialized for restructuring data from XML documents, rather than for general-purpose programming, though it is capable of some fairly powerful computational tasks. It is very different from other languages you are likely to have encountered. Non-programmers take heart: even if you had some other programming experience, it is not likely to help much with XSLT programming. Knowing HTML, on the other hand, is good background, since it is similar to XML.

This situation may seem a bit confusing at first: an XSLT stylesheet takes XML for input, often creates XML as output, and is itself XML. To help address this potential confusion, the XSLT language uses an idea called 'name spaces'; any XML tag that is actually a part of the stylesheet, as opposed to input or output, is marked as belonging to the XSL namespace. This is done by prefixing the tag name with 'xsl' and a colon. You will notice that most of the tag names in the stylesheet of (*Figure 3(b)*) start with 'xsl:'.

As an example, we will do a search on PubMed (see Chapter 1) and request the results in XML format (see Protocol 4). Now we can apply an XSLT stylesheet to rearrange this data into HTML format so we can put it on the web, or we can use a different stylesheet to format it in Rich Text Format (RTF), so we can open it in Microsoft Word.

Protocol 4

Retreiving PubMed Data in XML Format

1   Do your search as usual (see Chapter 1).

2   Click the checkboxes next to the articles for which you want to retreive the XML formatted bibliographic information.

3 Choose 'XML' on the display options pull-down menu (this is the menu with options for 'Summary', 'Brief', 'Abstract', etc.-'XML' is on this list).

4 Click 'Save'. You will be prompted for a file name, and you should give it an extension of '.xml'.

3.2 Example 3: Formatting PubMed results using XSLT stylesheets

For our first example of using an XSLT stylesheet, we will transform an XML document containing bibliographic references from PubMed. We will apply two different stylesheets to format this data in different ways. The first will create a web page, converting the XML input to HTML output. The second converts the references to a RTF document, so we can open it in a word processor such as Microsoft Word. In both cases we will apply typical formatting, such as making journal titles italic, volume numbers bold, etc., and we will put the information in an order closer to what journals expect.

The first step, of course, is to obtain PubMed results in XML format: this process is described in *Protocol 4*, which is basically a matter of making the appropriate choices on the NCBI PubMed web page.

For this example, the PubMed search was for articles using the term 'XML'. Five articles were selected, and the results were saved in XML format in the file 'XMLlit.xml'. The beginning of this file is shown in *Figure 3(a)*. Since XML data from the PubMed website is not consistently indented, the first citation has been formatted by hand, inserting indentation and line breaks to make the structure more clear. Indentation is for the benefit of human readers only, and is not necessary for the XSLT processor. Tree-structured data is hierarchical, in that elements can contain other elements. For example, the root element of our literature search is a 'PubmedArticleSet', which, logically enough, consists of a set of PubmedArticle items. These can be seen on lines 2 and 3 of *Figure 3(a)*. A PubmedArticle in turn contains a MedlineCitation (line 4), which contains (among other things) an Article element, which contains a Journal element, an ArticleTitle, an AuthorList, etc.

Understanding how the information elements relate to one another is key to manipulating and extracting XML data. Consider, for example, the Journal element (*Figure 3(a)*, lines 13–23). It contains two direct 'children', an ISSN (which pertains to the Journal in general), and a JournalIssue (which, of course, contains information related to a particular issue). The JournalIssue contains Volume, Issue number, and PubDate elements; the date in turn has a Year and a Month. Notice also that the AuthorList (which starts on line 32) contains multiple Authors, each of whom has a LastName, a ForeName, and Initials. This structure, a hierarchy of elements containing other elements, gives us a convenient way to identify particular elements in the XML file.

Our first stylesheet (see *Figure 3(b)*) converts the PubMed results to a HTML page. Note that the first line is the same as the first line of the PubMed data file in *Figure 3(a)*: both are XML documents adhering to version 1.0 of the specification. Line 2 is the opening tag for the `stylesheet` element, which is the root of the stylesheet. The stylesheet version is also 1.0, referring in this case to the XSLT specification. The last attribute of the `stylesheet` tag declares an XML namespace ('`xmlns`') called '`xsl`'; the namespace is associated with a URL, but only to ensure that the namespace is unique; the program does not actually contact that URL. Line 3 sets some output options; it tells the processor that we are constructing HTML output, and we would like it to be indented to reflect the nesting of the HTML elements. The rest of the stylesheet consists of four 'templates' that specify how data elements of the input file are to be processed. The important thing to notice about these templates is that they can be used in a nested manner that reflects the nested hierarchy of the elements in an XML document. The templates that handle higher-level elements call other templates to handle lower-level elements, which in turn may call more templates for still lower levels, until all desired levels are handled. This should be made clear by studying the templates in *Figure 3(b)*.

The first template (lines 4–14) applies to the root element, the `PubmedArticleSet`. Its match pattern begins with a forward slash to indicate that this is the root of the document (you may recognize that this use of slashes is similar to the Unix file system or to URLs on the web, which are similarly hierarchical). The contents of this template show what should be added to the output stream when the root element is encountered; lines 5–13 are simply the HTML code that will wrap up all the information. Nestled within this HTML framework on line 10 is a call to '`xsl:apply-templates`'; since this tag is in the xsl namespace, the processor recognizes it as a command, so it is not just blindly copied to the output. At this point the processor suspends its printing of the HTML from the root template and follows the instruction to process all of the `MedlineCitation` elements in all of the `PubmedArticle` elements. This processing will be done according to instructions from other templates in the stylesheet. When that is done, it will come back to outputting the HTML on lines 11–13, which will finish the job.

The next template (lines 15–35) matches `MedlineCitation` elements. Each will start by outputting an HTML list item (<LI>) tag, but then this template is suspended and another is called to deal with the article's list of authors. Once that is finished, execution of this template resumes. It pulls out individual items of information in the desired order, and wraps them in the appropriate HTML markup tags. These information items can be pulled from the current `MedlineCitation` using `xsl:value-of` tags, which each select an item using its 'path'. For example, '`Article/ArticleTitle`' refers to the `ArticleTitle` element of the `Article` element within the context of the current `MedlineCitation`.

For each article, we want to add a link back to PubMed so we can pull up the article's abstract, and, if we're lucky, the full text. This means we need to add an anchor tag to the HTML output with an HREF attribute containing a URL that will search PubMed for this particular article. Generating this sort of URL is described above in the CGI example (and also in Chapter 7). Most of this URL search string is the same for any article we want to find (see lines 27 and 28); the PubMed ID number at the end specifies the exact article we want, and is pulled from the input file in line 29. The `xsl:attribute` function lets us stuff this information inside the anchor tag.

The third template handles the lists of authors for each article. We could have used the same approach we did above by writing another template to handle an `Author`, then applying that template within the `AuthorList`. Instead, just to be different, we use a loop to successively indicate each `Author`, and print the relevant items of information. Some readers may find this looping approach more intuitive than nested template calls.

The fourth template (line 45) is a catch-all that tells the processor how to deal with elements that don't otherwise match a template (its select pattern is an asterisk, which is a wild card matching anything). Here we just ignore them, so the default template is empty.

To actually apply the stylesheet from *Figure 2(b)* to the XML data of *Figure 2(a)*, we need an XSLT processing program (see Web Resources). Using the saxon processor on Windows, the following command will do the trick:

```
saxon XMLlit.xml biblio_htm.xsl > XMLlit.htm
```

Figure 3(c) shows the resulting web page viewed in a browser.

The stylesheet shown in *Figure 3(d)* converts the same PubMed data to a different format. It is conceptually similar to the one discussed above, so we'll just consider the differences. Most importantly, the output of this stylesheet is in RTF, a widely used markup language for word processing documents. Rather than the angled brackets and tags of HTML, RTF uses notation such as `{\b bold}` and `{\i italic}` to represent **bold** and *italic*.

A more minor difference is that the reference numbers in this case are actually embedded in the output document (in the HTML version, the numbers were automatically supplied for the ordered list). An additional template has been added, which uses the XSLT `position()` function to find the number of each article in the set.

Now to view our PubMed data in RTF format, we simply apply our new stylesheet:

```
saxon XMLlit.xml biblio_rtf.xsl > XMLlit.rtf
```

Figure 3(e) shows the resulting RTF file opened in Microsoft Word. All or part of this document can be cut and pasted into other Word documents.

Please note that for serious page layout, including RTF or PDF output, it is not recommended to add the markup directly as we have done. An XML formatting language called XSL-FO ('XML Stylesheet Language Formatting Objects') is designed for this purpose; an XSLT stylesheet can be used to apply XSL-FO markup to a document, which can then be converted to the final format using an

XSL-FO processor. This approach is more robust and flexible that adding final format markup directly. More information can be obtained through the sites listed in the Web References, or in the books listed in the References (10).

3.3 Example 4: Converting BLAST output to a VRML model

BLAST (see Chapter 2) is a core tool for bioinformatics, and parsing BLAST results is a traditional chore that many people have done over the years in Perl (this has been described as a 'rite of passage' for bioinformaticians). The task of parsing is essentially writing a program to recognize the parts of the input data that fit certain criteria. Recent books address how to use Perl to parse this sort of biological data (7). Although a lot of people have done it, writing your own BLAST parser is not recommended. First of all, it is not trivial; the traditional BLAST output was designed to be (more or less) human readable, not to be read by a machine. It is easy to make errors, and a parser that is based on patterns in text is likely to be extremely sensitive to changes in the data format, so next time NCBI adds a new feature to their BLAST output, everybody's home-made parser becomes obsolete. For these reasons, parsing biological data is one of the central goals of the BioPerl (and BioJava, etc.) modules (see 'Creating a Bioinformatics Nation' in Web References).

NCBI's website (see Web Resources and Chapter 2) can return BLAST results in XML format. Since XML is a format designed to be read by computers, it should come as no surprise that it obviates many of the issues encountered by parsers of plain text files. If you are interested in filtering and/or rearranging the data from a BLAST search, you may be able to do what you need with an XSLT stylesheet rather than writing a parsing program. Here we will use an XSLT stylesheet to perform a fairly dramatic, if not very complex, transformation: to create a 3D VRML model (see Chapter 10) to visualize the results from a BLAST search.

Our example search uses the amino acid sequence of the *Taq* DNA polymerase (an enzyme commonly used for the Polymerase Chain Reaction) as a query against the known protein sequences. A small portion of the XML file returned from the NCBI website is shown in *Figure 4(a)*.

The root node of a BLAST output file is called, cleverly enough, `BlastOutput`. Note that this is NOT shown in *Figure 4(a)*, which shows only a portion of the file, starting at line 23. A node called `Iteration` within the `BlastOutput_iterations` node contains the results of our search. BLAST is capable of running multiple iterations of a search, using results from one pass to modify the search in the next pass. Since our search was of the old-fashioned, one pass type, there is only one `Iteration` in our results. Each iteration includes a set of `Hit` nodes, which represent genes with a significant match score to the query. Each gene has one or more high-scoring pair (`Hsp`) nodes, which represent regions of the hit gene that are significantly similar to the query sequence. Information reported about an `Hsp` includes the numbers of the first and last bases in the query and in the hit that match, various ways of reporting the score, etc. The

(a)

```
 TaqBLAST.xml                                                          _|□|X|
23      <BlastOutput_iterations>
24        <Iteration>
25          <Iteration_iter-num>1</Iteration_iter-num>
26          <Iteration_hits>
27            <Hit>
28              <Hit_num>1</Hit_num>
29              <Hit_id>gi|118828|sp|P19821|DPO1_THEAQ</Hit_id>
30              <Hit_def>DNA POLYMERASE I, THERMOSTABLE (TAQ POLYMERASE 1
31              <Hit_accession>P19821</Hit_accession>
32              <Hit_len>832</Hit_len>
33              <Hit_hsps>
34                <Hsp>
35                  <Hsp_num>1</Hsp_num>
36                  <Hsp_bit-score>1447.18</Hsp_bit-score>
37                  <Hsp_score>3745</Hsp_score>
38                  <Hsp_evalue>0</Hsp_evalue>
39                  <Hsp_query-from>1</Hsp_query-from>
40                  <Hsp_query-to>832</Hsp_query-to>
41                  <Hsp_hit-from>1</Hsp_hit-from>
42                  <Hsp_hit-to>832</Hsp_hit-to>
43                  <Hsp_pattern-from>0</Hsp_pattern-from>
44                  <Hsp_pattern-to>0</Hsp_pattern-to>
45                  <Hsp_query-frame>1</Hsp_query-frame>
46                  <Hsp_hit-frame>1</Hsp_hit-frame>
47                  <Hsp_identity>832</Hsp_identity>
48                  <Hsp_positive>832</Hsp_positive>
49                  <Hsp_gaps>0</Hsp_gaps>
50                  <Hsp_align-len>832</Hsp_align-len>
51                  <Hsp_density>0</Hsp_density>
52                  <Hsp_qseq>MRGMLPLFEPKGRVLLVDGHHLAYRTFHALKGLTTSRGEPVQA
53                  <Hsp_hseq>MRGMLPLFEPKGRVLLVDGHHLAYRTFHALKGLTTSRGEPVQA
54                  <Hsp_midline>MRGMLPLFEPKGRVLLVDGHHLAYRTFHALKGLTTSRGEP
55                </Hsp>
56              </Hit_hsps>
57            </Hit>
58            <Hit>
59              <Hit_num>2</Hit_num>
60              <Hit_id>gi|2506365|sp|P80194|DPO1_THECA</Hit_id>
```

(b)

```
 BLAST2vrml.xsl                                                        _|□|X|
1  <?xml version="1.0"?>
2  <xsl:stylesheet version="1.0" xmlns:xsl="http://www.w3.org/1999/XSL/Transform">
3  <xsl:output method="text" encoding="UTF-8" indent="no" />
4  <xsl:variable name="BAR_SPACING">12</xsl:variable>
5  <xsl:template match="*"/>
6  <xsl:template match="/BlastOutput">#VRML V2.0 utf8
7  Background { skyColor [0.9 0.9 1] }
8  Transform {
9      translation 0 0 0
10     rotation 0 0 0 0
11     children [
12 <xsl:apply-templates select="BlastOutput_iterations/Iteration/Iteration_hits/Hit"/>
13     ]
14  }
15  </xsl:template>
16  <xsl:template match="Hit">
17  # Hit number <xsl:value-of select="Hit_num"/>, accession <xsl:value-of select="Hit_accession"/>
18  Anchor {
19      url <![CDATA["http://www3.ncbi.nlm.nih.gov/htbin-post/Entrez/query?db=1&form=1&term=]]>
20  <xsl:value-of select="Hit_accession"/><![CDATA[+%5BACCN%5D"]]>
21      description "<xsl:value-of select="Hit_def"/>"
22      children [
23  <xsl:apply-templates select="Hit_hsps/Hsp"/>
24      ]
25  }
26  </xsl:template>
27  <xsl:template match="Hsp">
28      <xsl:variable name="similarity" select="number(Hsp_identity) div number(Hsp_align-len)"/>
29      Transform {
30          translation
31              0   # x
32              <xsl:value-of select="number(- $BAR_SPACING * number(../../Hit_num))"/> # y
33              0   # z
34          children Shape {
35              appearance Appearance {
36                  material Material {
37                      diffuseColor
38  <xsl:value-of select="number(1 - $similarity)"/>   # red
39  <xsl:value-of select="number($similarity)"/>   # green
40  <xsl:value-of select="number($similarity)"/>   # blue
41                  }
42              }
43              geometry Extrusion {
44                  crossSection [
45                      4 1, 4 -1, -4 -1, -4 1, 4 1]
46                  spine [ <xsl:value-of select="number(Hsp_query-from)"/> 0 0 ,
47                          <xsl:value-of select="number(Hsp_query-to)"/> 0 0
48                  ]
49                  solid TRUE
50              }
51          }
52      }
53  </xsl:template>
54  </xsl:stylesheet>
```

Figure 4 (*Continued*)

(c)

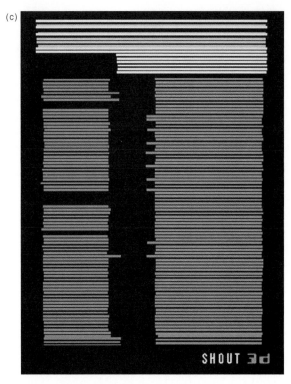

(d)

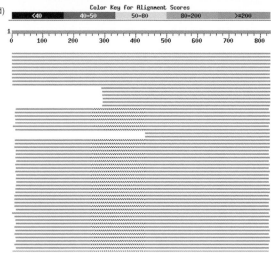

Figure 4 (a) A section of a BLAST output file as returned from the NCBI website in XML format. (b) Stylesheet to transform BLAST results into a VRML model. (c) VRML model of BLAST results, showing 100 hit sequences. This model is viewed from 'above', using the Shout3D Java applet. (d) The image map produced for the same data by the NCBI website. Only the first 50 sequences are shown.

actual sequences of the query and the hit are reported as well, along with the 'midline' sequence which is a sort of consensus between these two sequences. Note that there may be multiple `Hsps` in a `Hit`, representing different regions of a gene that match the query. For example, a gene might have two regions that each match the query, even though there is non-matching sequence in-between; this situation results in two `Hsps` in a single `Hit`.

Before we dive into the code of the stylesheet to transform this XML data, let's look at the output VRML model. Though VRML models simulate three dimensions, we will, for simplicity, only draw in a single plane, so that a ribbon extruded across the plane appears as a rectangle when viewed from above. *Figure 4(c)* shows the resulting model, which bears a distinct resemblance to the image produced on the NCBI website (*Figure 4(d)*), except that the colour scheme is different, and there are 100 sequences in our model, compared with only the first 50 in the image from NCBI. A bar representing the query sequence is shown across the top of the image, and each high scoring sequence segment discovered in the database is shown as a coloured bar below the query sequence. This type of figure reveals which portions of the query match parts of sequences from the database. Each database sequence is colour-coded to show it's overall similarity to the query (in the grayscale figure, lighter bars mean higher scores, though our model actually uses gradations from red to greenish-blue).

The NCBI web page uses the image as an image map, so pointing to a coloured bar brings up a message describing that sequence, and clicking on it takes you to the alignment of that sequence versus the query. In our VRML model, the bars are linked to the NCBI website.

Now let's go back to *Figure 4(b)* to see how a stylesheet can be used to create this sort of VRML model from our XML-formatted BLAST results. (This stylesheet contains a lot of VRML code, so it may be useful to review Chapter 10 first.)

There are three templates in the stylesheet that actually produce output. The first (lines 6–15) matches the root node. It generates the VRML label at the top of the file, sets the background colour, and sets up the outermost `Transform` node that will hold the parts to represent the various `Hsps`. The template encoded in lines 16–26 is called for each `Hit`. It creates an `Anchor` node, which lets us click on the contents of that node to navigate to a different website. These links take us to the `Hit`'s entry in GenBank. A third template (lines 27–53) is called for each `Hsp`, which will be represented by a rectangular bar made by extruding a box along the *x*-axis. The bar is enclosed in a Transform node to set its position in space. The colour of the bar is determined by its similarity to the query. Since we have to set red, green, and blue intensities to specify a colour, we calculate the similarity value once and remember it in a variable. This value is determined in line 28 (here, similarity is the number of identical residues divided by alignment-length).

In addition to these three output-generating templates, we have a default template (line 5). This is superficially different from the ones used in *Example 3* because it uses the abbreviated form of an empty tag. In XML, a tag with no contents, like <A></A> can be abbreviated to <A/>. But in both cases, they serve the same purpose, that is, to ignore elements for which there is no specific template match.

233

Finally, we have added one new fancy feature, a variable called BAR_SPACING (line 4). This controls how far apart the bars will be placed. This provides a simple way to change the model, by changing a single variable easily found at the top of the stylesheet, and running the data through the stylesheet again.

Each `Hit` is shown at a different *y*-position below the query. If a `Hit` has multiple `Hsps`, they are shown as segments at this same level. In the case of *Taq* polymerase used for this example, the `Hsps` represent the polymerase domain (on the right), and the $5'$-$3'$ exonuclease domain (on the left).

Saving this stylesheet in a file named 'BLAST2vrml.xsl' we apply it to the data using the Saxon XSLT processor as in the previous examples:

```
saxon TaqBLAST.xml BLAST2vrml.xsl > TaqBLAST.wrl
```

The third dimension is not actually used in these models, but it is interesting to think about how it might be exploited. For example, one might use the *z*-axis to show a sliding average of similarity to the query over a 'window' of several amino acids. Regions matching the query well would form 'peaks', and regions matching less well would form 'valleys'. Use of three dimensions along with more traditional effects like colour coding makes it possible to pack a large amount of data in a concise model.

3.4 Example 5: Handling XML data in Perl

Now that we have seen some examples of how XML and Perl can each be used in biological applications, we'll consider how they can be used together; specifically, how to manipulate structured XML data within a Perl program.

The XML::XPath module (available from Comprehensive Perl Archive Network (CPAN)) brings the same node-specfication syntax used in XSLT to Perl scripts, as shown in *Figure 5(a)*. This example performs a task similar to the stylesheets of *Figure 3*; it formats PubMed results, though in this case we simply output plain text. Lines 5 and 6 read XML data from the standard input and collect it into a single character string represented by the '`$article`' variable '$xml'. This data is used to construct a new instance of an XPath object on line 8. The rest of the program may then ask this object (or child objects derived from it) to find particular items of information as specified by XPath queries. For example, every article in our PubMed data file is represented by a MedlineCitation element. We ask the XPath object for a list of such elements, and process them one at a time in the loop represented by lines 11 to 29. Each time through the loop, another article is indicated by the '$article' variable. We can ask this article object for various information, such as its journal, title, volume, and so on. These simple attributes are collected as strings of characters (lines 19–25). Authors are more complicated, since there are different numbers of authors for different articles. Thus we need an inner loop (lines 14–18) to go through the list of authors for each article, and collect them together into a Perl list variable. Some of the formatting is done at this point, that is, last name, followed by a blank space, followed by initials. This list is converted to a character string on line 26, using

```
PubMedFormatter.pl                                                    _ □ ×
1  #!perl
2  use strict;
3  use XML::XPath;
4
5  my @xml = (<>);
6  my $xml = join("\n",@xml);
7
8  my $xp = XML::XPath->new(xml => $xml);
9
10 my $number = 0;
11 foreach my $article ($xp->find('//MedlineCitation')->get_nodelist){
12     $number++;
13     my @author_list;
14     foreach my $author_node ($article->find('.//Author')->get_nodelist){
15         my $author_name = $author_node->find('LastName')->string_value;
16         $author_name .= ' ' . $author_node->find('Initials')->string_value;
17         push @author_list, $author_name;
18     }
19     my $journal = $article->find('.//MedlineTA')->string_value;
20     my $title = $article->find('.//ArticleTitle')->string_value;
21     my $volume = $article->find('.//Volume')->string_value;
22     my $issue = $article->find('.//Issue')->string_value;
23     my $pages = $article->find('.//MedlinePgn')->string_value;
24     my $date = $article->find('.//PubDate/Year')->string_value;
25     my $pmid = $article->find('.//PMID')->string_value;
26     my $authors = join(", ",@author_list);
27     print   "$number. $authors. $title $journal. $volume($issue): $pages, ",
28             "$date. Pubmed: $pmid\n";
29 }
```

Figure 5 Perl script to format PubMed XML data as plain text.

the join() function to use a comma and a space to separate each author from the next. Finally, the parts are joined together with appropriate punctuation and sent to the output. Most of the formatting, such as determining the order of the parts or putting parentheses around the issue number, is done in lines 27 and 28.

There are many other Perl modules that facilitate the manipulation and generation of XML from within Perl (11), including those for the 'standard' XML Application Programming Interfaces (APIs) called Simple API for XML ('SAX') and the XML Document Object Model ('DOM'). These APIs describe XML documents and the operations that can be performed upon them in a standard vocabulary, which can be used in a similar way in programs written in Perl, Java, C++, or other languages.

4 Conclusion

The Internet has contributed in many ways to the feeling of information overload that many of us experience on a regular basis. Computers can store, communicate, and in some cases generate data at astonishing rates. When huge numbers of computers all over the planet (and beyond) are networked and internetworked, the pace of data production is multiplied even further. We humble humans may quite naturally feel overwhelmed. But our obvious response to the challenge of computer-aided data inundation is straightforward; we must use computers to deal with this data. We must develop skills to extract, structure, rearrange, and visualize data so that we may draw from it meaningful information and useful knowledge. Computers are essential tools for these tasks. It was not long ago in historical terms that computer users were invariably computer programmers.

Modern user interfaces have made it much easier for humans to use computers without having to program them. Nevertheless, simple programming skills can be valuable implements in the toolbox of people who use computers extensively. Molecular biologists increasingly fall into this category.

Web resources

BioPerl	`bio.perl.org`	A collection of software libraries for developers of Perl-based life-science tools
CPAN	`www.cpan.org`	Perl modules, scripts, and documentation
ActiveState	`www.activestate.com`	Perl (and other languages) for Windows and Linux
MacPerl	`www.macperl.com`	For Macintosh OS 7.0 and later. Note that MacPerl runs as a Classic environment on OS X, and that OS X comes with a native version of Perl that runs from the Unix prompt
2D illustration of proposed relationship of Prostaglandin H2 synthase to membrane	`blanco.biomol.uci.edu/ MolGraph/1prh.html`	From the 'Membrane Proteins of Known 3D Structure' collection of the Stephen White Laboratory at the University of California at Irvine
'Average' composition of a protein sequence	`ca.expasy.org/tools/pscale/ A.A.composition.html`	
SVG	`www.adobe.com/svg/viewer/ install/main.html`	This is the download site for the Adobe SVG Viewer, a web browser plugin. It has links to more information
ChemML	`www.xml-cml.org`	Follow the "information" link for white papers and other references

MathML	`www.w3.org/Math`	Includes links to tools and tutorials
MAGE-ML	`www.mged.org/Workgroups/MAGE/mage-ml.html`	A proposed standard data exchange format for microarray expression experiments
XSLT Processors		
Saxon	`saxon.sourceforge.net`	A version called 'Instant Saxon' is particualrly convenient to install on Windows machines. It uses the Microsoft Java Virtual Machine (JVM) that came with Internet Explorer, so you do not have to bother with installing and configuring a JVM
Xalan	`xml.apache.org`	Cross-platform XSLT processor (written in Java)
NCBI Data in XML	`ftp://ftp.ncbi.nih.gov/toolbox/xml/ncbixml.txt`	Comparison of the ASN.1 format used to structure most NCBI data with XML. Describes how one form is transliterated to the other
XSL-FO overview	`www.xml.com/pub/a/2002/03/20/xsl-fo.html`	We were able to do the RTF markup in Example 3 by incorporating RTF commands directly in our stylesheet because our formatting requirements were simple. For complicated RTF formatting, it is highly recommended to use more sophisticated approaches such as this

Creating a Bioinformatics Nation	`www.nature.com/nature/` `links/020509/020509-2.html`	Dr Lincoln Stein's article promoting a web services model for bioinformatics resources
Beginning Perl for Bioinformatics, Chapter 10 , GenBank	`http://www.oreilly.com/` `catalog/begperlbio/` `chapter/ch10.html`	Parsing GenBank files
NCBI BLAST Home Page	`www.ncbi.nlm.nih.gov/` `BLAST/`	See Chapter 2 for more information
XSL	`www.w3.org/Style/XSL/`	Overview and links to the specifications
Free On-Line dictionary of computing	`wombat.doc.ic.ac.uk/` `foldoc`	See entries for 'Perl', 'CGI', 'scripting', etc.

References

1. Filby, G. (ed.) (1997). *Spreadsheets in science and engineering.* Springer Verlag, Berlin.
2. Gibas, C. and Jambeck, P. (2001). *Developing bioinformatics computer skills.* O'Reilly & Associates, Inc., Sebastopol, CA, USA.
3. Rognes, T. and Seeberg, E. (2000). Six-fold speed-up of Smith–Waterman sequence database searches using parallel processing on common microprocessors. *Bioinformatics*, **16**(8), 699–706.
4. Wall, L., Christiansen, T., and Orwant, J. (2000). *Programming Perl*, 3rd edn. O'Reilly & Associates, Inc., Sebastopol, CA, USA.
5. Christiansen, T. and Torkington, N. (1998). *Perl cookbook.* O'Reilly & Associates, Inc., Sebastopol, CA, USA.
6. Horton, R. M. (1999). Biological sequence analysis using regular expressions. *BioTechniques*, **27**, 76-78.
7. Tisdall, J. (2001). *Beginning Perl for bioinformatics.* O'Reilly & Associates, Inc., Sebastopol, CA, USA.
8. Picot, D. and Garavito, R. M. (1994). Prostaglandin H synthase: implications for membrane structure. *FEBS Lett.*, **346**, 2125.
9. Stein, L. (1998). *Official guide to programming with CGI.pm.* Wiley Computer Publishing, New York.
10. Tidwell, D. (2001). *XSLT.* O'Reilly & Associates, Inc., Sebastopol, CA.
11. Ray, E. T. and McIntosh, J. (2002). *Perl & XML.* O'Reilly & Associates, Sebastopol, CA, USA.

Index